Your Successful Farm Business

Production, Profit, Pleasure

Your Successful Farm Business:
Production, Profit, Pleasure

by Joel Salatin

Polyface, Inc.
Swoope, Virginia

This publication is designed to provide accurate and authoritative information in regard to the subject matter covered. It is sold with the understanding that the publisher is not engaged in rendering legal, accounting or other professional service. If legal advice or other expert assistance is required, the services of a competent professional person should be sought. *From a declaration of principles jointly adopted by a committee of the American Bar Association and a committee of publishers.*

Your Successful Farm Business:
Production, Profit, Pleasure, First Edition
Copyright © 2017 Joel Salatin

Editing and book design by Jennifer Dehoff

The Cover: Artwork by Rachel Salatin

Library of Congress Control Number: 2017937554

ISBN: 978-0-9638109-8-4

Other Books By Joel Salatin

The Marvelous Pigness of Pigs:
Nurturing and Caring for All God's Creation

Growing up straddling the tension between the environmental and faith-based community, several years ago Joel poked good-naturedly at the stereotypes with his self-acclaimed moniker: Christian libertarian environmentalist capitalist lunatic farmer. Friends in both camps have marveled at how he could be so ecological but yet read the Bible. Aren't the two mutually exclusive? The thesis of the book is simple: All physical creation is an object lesson of spiritual truth. The question is simple: Do the beliefs in the pew align with what's on the menu?

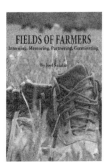

Fields of Farmers:
Interning, Mentoring, Partnering, Germinating

America's average farmer is sixty years old. When young people can't get in, old people can't get out. Approaching a watershed moment, our culture desperately needs a generational transfer of millions of farm acres facing abandonment, development, or amalgamation into ever-larger holdings. Based on his decades of experience with interns and multigenerational partnerships at Polyface Farm, farmer and author Joel Salatin digs deep into the problems and solutions surrounding this land and knowledge-transfer crisis. This book empowers aspiring young farmers, midlife farmers, and nonfarming landlords to build regenerative, profitable agricultural enterprises.

Folks, This Ain't Normal:
A Farmer's Advice for Happier Hens, Healthier People, and a Better World

From Joel Salatin's point of view, life in the 21st century just ain't normal. He discusses how far removed we are from the simple, sustainable joy that comes from living close to the land and the people we love. Salatin has many thoughts on what normal is and shares practical and philosophical ideas for changing our lives in small ways that have big impact. Salatin understands what food should be: Wholesome, seasonal, raised naturally, procured locally, prepared lovingly, and eaten with a profound reverence for the circle of life. And his message doesn't stop there. Salatin writes with a wicked sense of humor and true storyteller's knack for the revealing anecdote.

Book list continues on next page.

Other Books By Joel Salatin

Holy Cows and Hog Heaven:
The Food Buyer's Guide to Farm Friendly Food
Written for food buyers to empower them in their dedication to food with integrity, this book changes people's lives. Farmers who give it to their customers say that folks who have read it have a new level of understanding and a delightful attitude about the farmer-consumer relationship. A short, easy read, this book will make you laugh and cry, all in a matter of minutes. Any consumer wanting to peek in to the life behind the local food farmer will be delighted at the insights and real-life stories Joel shares from his own marketing experience.

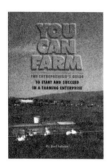

You Can Farm: The Entrepreneur's Guide to Start and $ucceed in A Farming Enterprise
For all the wannabes and newbies. A veritable compendium of information, Joel pulls from his eclectic sphere of knowledge, combines it with a half century of farming experience, and covers as many topics as he can think of that will affect the success of a farming venture. He offers his 10 best picks for profitable ventures, and the 10 worst. He covers insurance, record keeping, land acquisition, and equipment. A hard hitting, practical view from a successful farmer, if this book scares you off, it will be the best reality check you ever spent.

$alad Bar Beef
Fishing for a phrase to describe this ultimately land-healing and nutrition-escalating production model, Joel realized that he was offering the cows a salad bar. He coined the phrase to describe the farm's beef, and thereby stimulate questions from potential customers. This book describes herd effect, mobbing, moving, field design, water systems, manure monitoring, soil fertility, and even pigaerating. A fundamentally fresh way to look at the symbiosis between farmer, field, and cow, this book is now a classic in the pasture-based livestock movement.

Pastured Poultry Profit$:
Net $25,000 on 20 Acres in 6 Months
Joel began raising chickens when he was 10 years old and serendipitously fell into the pastured poultry concept a couple of years later. Still the centerpiece of the farm, and the engine that drives sales, notoriety, and profit, pastured poultry has revolutionized countless farming endeavors around the world. A how-to book, this includes all the stories and tips, from brooding to marketing. Centered around meat chickens, it includes a section on layers and turkeys. Many would say this book started the American pastured poultry movement.

Other Books By Joel Salatin

The Sheer Ecstasy of Being a Lunatic Farmer

Have you ever wondered: So what really is the difference, anyway? Can there really be that much difference between the way two farmers operate? After all, a cow is a cow and the land is the land, isn't it? Gleaning stories from his fifty years as localized, compost-fertilized, pasture-based, beyond organic farmer, Salatin explores the differences. From how farmers view soil and water, to how they build fences, market their products or involve their families, this book shows a depth of thought that expresses itself through farms like his family's Polyface Farm. In the international spotlight for this different kind of farm, Salatin explains a different food model and shows with good humor and stories how this alleged lunacy actually offers a life of sheer ecstasy.

Everything I Want to Do Is Illegal:
War Stories from the Local Food Front

Although Polyface Farm has been glowingly featured in countless national print and video media, it would not exist if the USDA and the Virginia Department of Agriculture and Consumer Services had their way. From a lifetime of noncompliance, frustration, humor, and passion come the behind-the-scenes real stories that have brought this little family farm into the forefront of the non-industrial food system. You may not agree with all of his conclusions, but this book will force you to think about things that most people didn't even know existed.

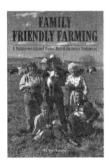

Family Friendly Farming: A Multi-Generational
Home-Based Business Testament

Few life circumstances are as hard to navigate as family business. Today, four generations of Salatins work and live on Polyface Farm. It's not easy, but this book describes the rules and relational principles to harmonize in what is too often a tense environment. The chapters on how to get your children to enjoy working with you are worth the price of the book. But beyond that, it delves into the quagmire of inheritance, family meetings, and personal responsibility. These are thorny situations, but a pathway exists to leverage the strengths of family business. The goal of this book is to hold families, and especially family farms, together.

Books Available From

Polyface Farm Gift Shop	1-540-885-3590	www.polyfacefarms.com
Chelsea Green Publishing	1-800-639-4099	www.chelseagreen.com
Acres USA Magazine	1-800-355-5313	www.acresusa.com
Amazon.com		www.amazon.com
Your local bookstore		

Joel Salatin DVDs

The POLYFACE PRIMER SERIES is a collaborative educational endeavor between Polyface Farm, operated by the Salatin family, and Swineherd Productions, operated by Shrader Thomas. Joel has written numerous books outlining the farm's techniques and philosophies, but now the family brings these principles into the living room and the classroom via video. In these episodes, you'll see different members of the Salatin family as well as interns, apprentices and staff.

Primer Series: Pigs 'n Glens

Pigs are omnivores, like humans. We don't use the word glens much anymore, but you'll find it in old fairy tales and fold lore with names like Rip Van Winkle and Ichabod Crane. It describes a forested setting in which the story will take place. This video combines the most ancient hog production techniques with the best of modern technology, using pigs to massage the ecological landscape in exercise. So hold onto your hats, here we go... Released June 2013 (40 minutes).

Primer Series: Techno Stealth: Metropolitan Buying Clubs

Over the years, we have developed a local food distribution system that we call the Metropolitan Buying Club. We think it combines the real-time interfaces of online marketing with community-based interaction. These kinds of interfaces, without bricks and mortar, using the internet, create efficiencies and economies of scale in local food distribution that we think you will find very exciting. Released July 2014 (45 minutes).

Polyface Farm

This is the official, comprehensive video about the farm. If a picture is worth a thousand words, this video is worth a library of writing. Carefully filmed and edited by Moonstar Films, this is a first class, professional work. Filmed over the course of a year to catch the multi-faceted seasons and enterprises at Polyface, this was a monumental effort and a delightful artistic piece. You will see Joel and Daniel in each section, talking and working in everyday settings. Released 1998 (110 minutes).

Polyfaces: A World of Many Choices

One Australian family spent 4 years documenting a style of farming that will help change the fate of humanity! Produced over 4 years it follows the Salatin's, a 4th generation farming family who do 'everything different' as they produce food in a way that works with nature, not against it. Using the symbiotic relationships of animals and their natural functions, they produce high quality, nutrient-dense products. Set amidst the stunning Shenandoah Valley in northern Virginia, 'Polyface Farm' is led by "the world's most innovative farmer" (TIME) and uses no chemicals and feeds over 6,000 families and many restaurants and food outlets within a 3 hour 'food-shed' of their farm. This model is being replicated throughout our global village, proving that we can provide quality produce without depleting our planet. Released 2015 (1 hour 32 minutes).

TABLE OF CONTENTS

Acknowledgments xiii

Foreword xiv

Introduction xvii

Chapter 1: **Working Landscapes** 2
    Domicile 10
    Access 11
    Fences 14
    Water 16
    Buildings 18
    Vegetation cutting 19

Chapter 2: **Eclectic Awareness** 22
    The local daily newspaper 29
    Tribal magazines 30
    Books 31
    The opposition 31

Chapter 3: **Live Frugally** 36

Chapter 4: **Can Do Entrepreneurial Spirit** 56
    The Fears: Knowledge 65
        Acquiring land 66
        Finances 67
        Labor 68
        Marketing 68
        Business 69
        Optimism 71

Chapter 5: **Assemble A Team** 72
    Starter vs. finisher 78
    Messy vs. cleany 78
    Spender vs. saver 79
    Introvert vs. extrovert 80
    Team Development: Micromanagement 84
    Team Development: Working hard vs. free time 85

Chapter 6:	**Direct Marketing – Why**	**92**
	Self promoting	94
	Emotionally vested	97
	Hard work	97
	Peer dependency	97
	Community economy	103
	Attracts the best and brightest	104
	Customers will move with you	105
	Emotional support	106
	Polyface in Photos	**108**
Chapter 7:	**Direct Marketing – Where**	**126**
	Big supermarkets	126
	Small independent grocers/supermarkets	130
	Food co-op	131
	Farmer's markets	132
	Electronic aggregators	135
	Restaurants	139
	Community Supported Agriculture (CSA)	141
	On-farm sales/store	143
	Metropolitan Buying Club (MBC)	145
	Food trucks, kitchens	148
Chapter 8:	**Direct Marketing – How**	**150**
	Diversified portfolio	150
	Differentiation	153
	Customer friendly	156
	Delivery is a separate business	159
	Everyone food	162
	Find your fit	163
	Cash flow	164
	Gateway products	165
Chapter 9:	**Gross Margin Analysis**	**168**
Chapter 10:	**Multi-Enterprise**	**186**
Chapter 11:	**Stay Nimble**	**202**
	Mobile farms	204
	Modular	213
	Management intensive	216

Chapter 12: **Time & Motion Studies** **224**
    Set benchmarks 231
    Go loaded and come loaded 239
    Make lists 244

Chapter 13: **Getting Started (or Starting Over)** **248**
    Mid-lifers 248
    Young professionals 254
    Youth and 20 somethings 256

Chapter 14: **Distractions** **262**
    Bank barns 262
    Horse 264
    Off-farm recreation 265
    Heritage genetics 266
    Certifications 268
    Government agencies 271
    Social media 272
    Altruistic addictions 273
    Unscheduled visitors 276
    Failure to cull 277
    PVC construction 279
    Cheap infrastructure 281

Chapter 15: **New Opportunities** **284**
    Agri-tourism 284
    Edu-tainment 287
    Urban custom farmers 289
    Agri-community 290
    Therapy farms 293
    Elder-care farms 296
    Farm schools 299
    Camp farms 301

            Summary **304**

            Index **308**

Acknowledgments

I stand on the shoulders of real geniuses, years of reading, and lots of personal conversations listening to the successes and failures of many farmers. Like most informational books, this is in many ways a condensed version of all those bits of relationships and stories.

As always, what makes my books possible is an amazing family and staff that keep the farm running to enable me to peck away at the laptop and crank this stuff out. First and foremost, thanks to my wife, Teresa, who keeps the strings of the household together and protects me from myself. We've been loving each other for 36 years.

This is the first book that showcases our grandchildren, Travis, Andrew, and Lauryn, which is exciting. All of them found and developed their own enterprises under the tutelage of Daniel and Sheri, our son and daughter-in-law.

As usual, the creative and artistic talents of our daughter, Rachel, developed the cover. Jennifer Dehoff's editing and desktop publishing made everything more readable.

All of the most significant mentors who developed my thinking early on have passed away--my dad, Allan Nation, Reid Putney, Charles Walters, Don and Ellie Pruess, A.P. Thomson, Robert Rodale--but their memory lives in these pages.

My mother, an extremely active and agile 93 year-old, definitely instilled the people skills and performance attributes that gave me storytelling genetics. I don't have much that wasn't given or borrowed from others.

Our Polyface farm staff keeps things humming and provides endless grist for the story mill--how to do it right, and how to do it wrong. I'm forever indebted to their loyalty and their "sharpening of my saw," in the words of Stephen Covey. I've been blessed beyond measure, and hope this distillate blesses you as well.

Foreword

Joel Salatin is the Pied Piper when it comes to pastures. People flock from all over the world to walk in his footsteps at Polyface Farm. They see the dozen-plus enterprises (the number constantly changes) and edge as close as they can to hear how it all works so that they can apply some of the lessons for their own operations.

They see the success. They see the innovation. They experience first-hand the sustainability of clean pastures and clean food.

What they don't see is what it took to get to the point where Polyface thrives today. Your Successful Farm Business gives insight into the frugal early years when Joel was known as The Chicken Man. From this back-story, through the struggle-and-grow years, the book leaves readers with food for thought about enterprises for the future.

In 1982 Joel and his wife, Teresa, quit off-farm jobs to follow their dream to be full-time farmers. Everyone thought they were crazy. They had saved enough money to give themselves a year to accomplish their goal.

Joel is quick to point out that from the beginning he had partners, first his wife and his parents, and they lived in the attic of his parents' home. He and Teresa got up at 4:00 AM to process chickens together, paused for breakfast, then processed more chickens so that by 1:00 PM they would be ready to receive customers who arrived to purchase chicken at the farm. By the end of the first year they had successfully managed a sustainable living from farming.

Your Successful Farm Business goes beyond You Can Farm, which he wrote 20 years earlier. He describes this book as "more advanced, more refined, more mature, simply more." Throughout, he makes references to his previous books for readers who want to delve deeper into the subjects briefly covered here. While writing this book he studiously made a point not to plow previous ground, instead offering fresh views of his beliefs and methods. Anyone who is determined, passionate about their dream, and willing to start out on a shoestring can achieve the same kind of success as Polyface, whether they just read about it or actually set foot on the farm.

Over the years Joel has changed his moniker of Chicken Man to that of Christian-libertarian-environmentalist-capitalist-lunatic-farmer to proclaim that he can't be put into a cookie-cutter box.

Joel writes that "orthodox industrial farming fails first because it fails to feed the human spirit." You will find plenty of Joel's spirit in Your Successful Farm Business. His Joelness shines through as he shows how to make a successful, sustainable, decent living farming, "while much of the world is fighting traffic in daily commutes to work in confined cubicles attached to electronic devices." His most passionate goal is to eliminate confinement farming and replace it with credible alternatives.

He believes the more you diversify the farm, the more you follow Nature's template. "A successful farm expresses a never ending quest toward better," he writes. "If you want to farm, come to it for love, for passion, for creation, stewardship, redemption."

In school Joel excelled in debate. He confesses that he's an extrovert who thrives on meeting, talking to and being with people. Polyface Farm offers 24/7/365 transparency to visitors. However, as much as his interaction with people energizes him, Joel realized he had to find a balance between passion and pragmatism in order to get the necessary chores done. As a result he instituted scheduled

Lunatic Tours to group visitors on organized outings. This has allowed breathing (and working) space while honoring his desire to share his triumphs and even some failures on the farm.

In demand as a speaker world wide (how could he turn down the invitation to speak to Allan Savory's London gathering of the Savory Institute, or meet Prince Charles after another speaking engagement?), he somehow manages to balance these on-the-road engagements while keeping the vision for Polyface on track. There is nothing static about Polyface Farm. It is constantly growing and changing.

Joel loves working with his hands, wearing work clothes, and sporting calloused hands. His favorite activity is working with a chainsaw. Those who meet and hear him speak often offer their mountain cabin or beach house as a getaway for him, but Joel says his getaway is the farm. It's his love. It's his life. And he delights in sharing all he can to show others how they can enjoy and profit by living the farm life too.

For all of his theatricality, his world wide fame, and his bold and oft times controversial challenges of the status quo, there is a true humility at the heart of his Joelness. That too is revealed through Your Successful Farm Business.

Joel confesses that he is "hopelessly in love with healing the land, serving enthusiastic clients with honest nutrient dense foods and promoting a farm and food ethic that honors farmers." That enthusiasm will rub off on readers who take Your Successful Farm Business to heart.

>   Carolyn Nation
>   Co-managing Editor
>   *The Stockman Grass Farmer*

Introduction

Twenty years ago I put everything I knew into a book titled *YOU CAN FARM*. At the time, I could not imagine a sequel. I thought I had said everything that could possibly occupy my little brain about the subject. But that was twenty years ago.

Since then, principles only hinted at have been fleshed out, refined, and developed. Things I could not have even thought of then now dominate my thinking. Several years ago Allan Savory asked me to speak at his international London gathering for the Savory Institute. How could I turn that down?

I pondered a lot about what to say to such a well-read and accomplished group. People who attend these types of conferences are not lightweights. They're heavy hitters, the leaders, the movers and shakers. It occurred to me that perhaps it would be good to try to articulate the common threads of success for beginning farmers. In my experience and travels, what did I see that made a farmer thrive? That particular perspective seemed like a worthy addition to present.

If we could concisely articulate these principles, perhaps it would help us to encourage the next generation of farmers following our lead. Kind of a "here's what I wish someone had told me when I was a starry-eyed wanna-be farmer." Don't all of us want to be 20 years old again . . . but as wise as we are at 60?   I pitched the idea to Allan and he thought it was a good fit.

That talk I gave in London was well received and is still one of my favorites and most requested. Often when I step off the stage

xvii

after delivering it, people come up to me and ask, "Have you written this down? This is good stuff. Can I get a copy of it?" Gradually it dawned on me that these principles were the graduate level of my earlier "tell all" attempt with *YOU CAN FARM*.

I decided I'd better write it down, and when I realized I was coming up on the twentieth anniversary of that earlier book, it seemed like good timing. I looked through *YOU CAN FARM* to make sure these new ideas were not in it. I couldn't believe I'd written 500 pages on this topic before and not put these concepts in it. Some had been mentioned only in elementary form. Some were not included at all. None was refined to the point you'll see in this book.

For some time I toyed with the title *YOU CAN FARM 2.0*. But then it dawned on me that such a title would appear to make the first *YOU CAN FARM* obsolete and unnecessary. Please understand, for an unpublicized self-published book, *YOU CAN FARM* has done beyond extremely well. And it's still plugging along, riding the self-reliance, prepper, homestead, agrarian self-employed wave. I didn't want to jeopardize its current sales or suggest the still valid information it contained isn't worthy of a read. In true marketing bounce-back metrics, I wanted to duplicate that effort without losing the earlier momentum.

I opted for the *YOUR SUCCESSFUL FARM BUSINESS* title to create separation and validate that each book can stand alone. Once a marketer, always a marketer. Make no mistake, *YOU CAN FARM* should be read before reading this book. It is the foundation, the high school diploma. *YOUR SUCCESSFUL FARM BUSINESS* will be far richer if you read the other one first. Just sayin'. The earlier work is certainly not outdated. It's absolutely current, and I have purposely steered away from any redundancy in order to add to the body of information rather than repeat it.

Of course, I hope everyone who ever read *YOU CAN FARM* will now want to read this new work. The guts of it are the original ten threads of beginning farmer success. I used that talk as my keynote address at the six *Mother Earth News* magazine fairs a couple of years ago. By the way, if you haven't been to one of those, you should go: they're fantastic.

New insights come at funny times. Many of these came as an epiphany while I was working in the field. Others came as a result of conversations with mentors like Allan Nation, founder and editor of *Stockman Grass Farmer*. Now that I've assumed the editorship of that publication, I'm hoping more strokes of genius will come my way. Some of these principles became clear in the angst of difficult discussions with struggling farmers, kind of a what could this farmer have done to keep from falling into the difficult situation?

Over time, over time, over time, if we're listening to instruction, conversing and reading widely, and translating that into experience in our own businesses, the synthesis takes up residency. When an idea translates to deed and you see it work or not work, that's validation. Over time, validated principles develop a body of active truth. I have done the best I know how, as honestly as I know how, as precisely as I know how, to convey these principles so that all of us will do better and aspire to more.

With the average farmer at 60 years old, and 50 percent of all agricultural assets due to change hands in the next 15 years, we are entering an unprecedented time of disruption in our farm-scape and our food system. It's time to seize the opportunity. This is not a time to put our face in our hands and weep for what was or what could have been.

It is a time to embrace the innovative lunatic fringe and walk boldly into an agriculture that is productive, profitable, and pleasurable. Will this unprecedented agrarian equity transfer go to foreign powers, corporate interests, and industrial farmers? Or will it go to a new generation of bright eyed, bushy-tailed self-starter entrepreneurs who will caress the land and its dependents? This is my dream and my hope. I pray it's yours as well.

Together, let's explore the principles that will make *YOUR SUCCESSFUL FARM BUSINESS* enjoy Production, Profit, and Pleasure.

Joel Salatin
Polyface Farm
Summer, 2017

Your Successful Farm Business

Production, Profit, Pleasure

Chapter 1

Working Landscapes

By definition, farmers interact intimately with the landscape. Why do we call a farm a farm? I ask this question routinely on my Lunatic Tours (scheduled hay wagon tours at our Polyface Farm) and the answers run a wide gamut. When I ask the question, the interchanges often go like this.

Guest 1: "It produces something."

Me: "Well, a park produces squirrels; a wilderness area produces trees."

Guest 2: "It produces food."

Me: "Isn't a squirrel food? Or the dandelions growing by the concrete walk by the strip mall?"

Guest 3: "It has animals."

Me: "You mean a farm that produces vegetables isn't a farm?" Everybody on the tour laughs good-naturedly.

Guest 4: "It's food and fiber to sell."

Me: "Oh, so if I go out to a wilderness area and find ginseng and sell it to the Chinese, then suddenly we call it a farm? How about if I harvest an elk, pay a taxidermist to mount the head, then sell the head at the local fire company fund-raiser--does that make the Colorado forest where I hunted a farm?"

By this time, people are really scratching their heads and the usual left-field remarks come in.

Guest 5: "It gets government subsidies."

Me: "Ha!"

Guest 6: "It doesn't have very many people on it."

Me: "You don't really mean that."

Guest 7: "It stinks!"

Me: "We're standing here in our field with 50 pigs right over there, a herd of cows behind us, and 2,000 chickens in front of us. Does it smell?"

A chorus of "No" erupts from the now thoroughly bewildered tour attendees.

Guest 8: "A tractor!"

Me, laughing: "So if I do my field work with horses, it's not a farm."

Guest 9: "A barn."

Me: "Well, you're getting closer, but if we put a barn in the Shenandoah National Park, would we suddenly call it a farm?"

Guest 10: "I've got it! I've got it! A FARMER!"

Me: "Move to the head of the class. Yes, that's it. A farm is a piece of land that a farmER touches, develops, and utilizes. The only difference between a farm and any other piece of real estate is the farmER."

Then I usually go into a short rant about the fallacy of assuming you can preserve farmERS if you preserve farmLAND. Without farmERS farmLAND simply becomes wilderness. In our part of the world, that means it gradually changes from pastures, fields, and gardens to brambles, bushes, and then trees. Most open land is borrowed from the forest.

And the reason America has a dynamic and active TREE FARM designation is so the forestry industry can recognize the difference between human touched, or stewarded forests, and those left untouched. The human component of touching the landscape is the distinguishing characteristic of a farm. Obviously humans touch the landscape all the time, from building skyscrapers to building dams on the Mississippi, but without a farmER, those activities likewise don't engender the landscape designation of FARM.

The way to preserve farmLAND is to preserve farmERS. This seems like a simple truism, but of course, the idea encompasses a complex set of issues. Markets, agronomy, husbandry, mechanics, infrastructure, technique--goodness, the list is virtually endless. The point is that a lot of things have to be in place in order to create a conducive habitat in which farmERS can thrive.

Perhaps no issue is more fundamental than the idea that farmers must interact with their landscape, to participate with its succession, to develop it into something that doesn't look like it would if it were abandoned.

And herein lies the crux of the successful thread: working landscapes.

Anyone who studies the relationship between humans and the environment knows that the history of this interaction more often than not bodes ill for the ecology. You don't have to be a tree hugger greenie-weenie to realize that the rise and fall of civilizations usually mirrors the rise and fall of resources. A culture can only be as wealthy as its air, soil, and water. But depletion and exploitation too often describe human stewardship.

Abundant production, whether gathered from the wild or produced from fertile fields, necessarily creates thriving civilizations. Populations increase, social structure develops, and commerce rises in choice and opportunity. Usually this trend extracts and depletes the resource base until the society collapses or gets overrun by competing societies that have figured out how to extract more resources faster. I don't want to oversimplify here, but I don't know anyone who would say today that the environmental human track record has produced more potable water, more breathable air, and more soil on the earth's surface.

The nearly universal long-term effect of human occupation has been ecological degradation and depletion of the common resource base, either fast or slow. The result is that most of us who care about the commons carry a giant burden on our back, a great weight of remorse. As we look back through the course of history and weep over what our ancestors did, the notion that we could interact properly seems impossible.

Our mechanical prowess and intellectual capacity have hurt the planet deeply, especially in recent times, and this makes us timid and gun-shy about touching the earth. This has led to a practically universal worldview, in what is broadly known as the environmental movement, of a foundational principle I call environmentalism by *abandonment.* This is an incredibly disempowering notion for a vocation or a person who by definition plans to interact with the ecology of a place.

I appreciate and understand the remorse, the burden, the fear. The mainline environmentalist, therefore, consistently advocates for policies that remove people from the land. Whether it's the idea of buffalo commons or easements that preclude day camps and intern housing in agricultural zoning, this abandonment notion permeates the mind of most people.

Look at almost any landscape battle and it's built on this abandonment foundation. How much wilderness area does the country need? Apparently at least one more acre . . . forever, if you look at the wilderness area advocacy agenda. When is enough, enough? State parks, national parks, national forests, sensitive designations--locking up land and getting rid of humans has almost become a national pasttime.

I appreciate this effort to protect land and understand why the abandonment mentality dominates our landscape and zoning policies. Along with that is a prejudice against farming as a filthy, stinky, unsightly business performed by hicks and intellectually-disadvantaged people. I get this too. Industrial farming has certainly stimulated the stereotype. How else can you account for farmers still waiting in line to mortgage their lives as voluntary serfs to build massive chicken houses for someone else with no guarantees that they'll get chickens to raise or a decent price per pound?

Most Americans would say that farmers in general do far more damage to the environment than suburban lawns and brick ranchers fronting on paved streets with storm drains, attached two-car garage and appropriate entertainment centers spread throughout. While the mainstream agriculture community accuses urban and suburban chemical lawn care and soap suds from pet washing as the biggest environmental hazard, the truth is that these urban activities did not create a dead zone the size of New Jersey in the Gulf of Mexico. The stench emanating from a factory livestock facility is proof enough. And soil loss? Desertification in the southwestern U.S. is increasing as fast as anywhere in the world. The magnificent grasslands of early ranges and cowboys now barely grows a piece of vegetation in a square yard. Bare soil is the new normal.

Regenerative farmers notwithstanding, the agricultural environmental trajectory does not lead to a good place. The environment-friendly farmers entering the field literally look at mainstream orthodoxy like a battle survivor surveying bodies and bomb craters. We can't believe what we've done to ourselves. We wonder where to start the clean-up. Triage screams one word: "Everything!"

We can flagellate ourselves at the proverbial penitence whipping post all day and it doesn't change the reality of the data. It doesn't bring back the soil. It doesn't refill the aquifers. It doesn't fix the fecal pall hanging over Colorado feedlots. And it certainly doesn't restore the honey bees and animal/plant diversity that built soil for centuries prior to America's heartland wealth being mined into corn and soybeans.

It's not a pretty picture, this human interaction, this modern farming system. Sir Albert Howard, godfather of modern aerobic composting, said that every generation is tempted to take what nature built in millenia--fertile soil--and turn it into cash. Nobody could say it better. And so this reluctance to touch, to interact with nature is ubiquitous.

The "leave no trace" rule at parks, recited by hikers everywhere, bolsters this "nature is pristine; humans are nasty" notion. In no way am I advocating that we should litter our parks or eliminate Yellowstone, but when it comes to the work of the farmer, the daily outworking of this vocational choice, we must recognize the societal and personal bias against altering the landscape. And yet this is exactly what the farmer does . . . in spades.

You can't farm without altering the landscape. If you plant one tomato or cut one tree limb, you've altered the landscape. As a result, farmers who succeed must completely change this cultural narrative at a deeply subconscious level. We must have a conversation with our inner selves that goes something like this: "I'm sorry and deeply regret all of the damage my ancestors have done to the earth. I'm sorry for what today's farmers, by and large, are continuing to do to the earth. But these hands that others have used to hurt can also be used to heal. And I am committed to reversing the Conquistador mentality, to bring nurturing and healing instead."

Beyond that, would-be farmers must believe, deeply, that we can alter the landscape in a way that not only yields an income, but grows the commons and yields more soil, more breathable air, and more potable water as a result of our interaction. Embracing this mentality frees us up--spiritually, emotionally, physically--to take on the landscape adjustment that is the signature of every farm.

Adjustments are not always pretty at the beginning. Goodness, the cast necessary to heal a broken bone isn't attractive. It's downright ugly and a nuisance. But it's part of the healing process. The house you live in didn't start with beautiful shade trees and the hammock on the back corner. It started out as an ugly backhoe trench for footers, perhaps with boards laid out across what is today a lawn, in order to keep from getting gooey, sticky mud on your shoes during the daily visits to the construction site.

Anything worth learning, worth building, and worth doing takes adjustments. Repentance, or asking forgiveness, is not pretty.

But it yields beauty in healed relationships and deep freedom. To assume that the way things are--either with a landscape or with my own thinking--is automatically the best is both childish and foolish. Life is all about adjusting--otherwise known as growth. Growth is adjustment, wouldn't you agree? Moving away from diapers is progress we can all get behind. That's a good adjustment.

To assume that every rock, every tree, every valley or hillside is in its absolute best position or design the way it fell after the last volcano, earthquake, or tectonic plate shift is simply ridiculous. And yet many folks look at nature with this level of untouchable faith. Lots of things alter the landscape: floods, tornadoes, fire, bison, prairie dogs. Change has been going on for a long time and will continue, even without human touch. We don't have to change everything, but change is part and parcel of daily activity on this planet.

Change can be positive or negative. But it's universally shocking. When Teresa and I attend our high school class reunions, sometimes we can't recognize classmates. We try to go to see how much everyone else has changed. Ha! Some people have died. That's a huge change. Some people are incredibly sick, and their triage stories break my heart. But at the end, I come away grateful that their caregiver cares enough to intervene and offer therapy or adjustments that move the person toward recuperation.

This is the way farmers must look at our land. We view it like a patient, a partner, a work-in-progress. If we don't, we'll be unable to aggressively interact in a redemptive way. The reason this is important is because many new farmers today come from an environmentalism paradigm. This is quite different than yesterday's farmers. A century ago farmers did not embrace farming because they wanted to heal anything. They embraced it because it was a way to earn a living.

Does the average person who chooses a career in informational technology come to it because she's trying to heal corporate America? Or does she first and foremost want to leave a legacy on the computer software industry? No. Most come to their work--their job--as simply gainful employment. It beats poverty and enables you to keep shoes on your feet. The current employment statistics that 80 percent of Americans hate their job indicates a profound disconnect between vocation and legacy.

I think when more than half of Americans were farmers they joined the vocation much the same way. It was a way to earn a living; it was interesting; people needed what they did; it attracted interesting people that formed an interesting support community... you get the drift. For many, of course, it was a way to express a dominion mandate, to carve out order in an otherwise chaotic and indeterminate ecology.

But today's new farmers are different. For the most part we're environmentally caring, socially committed, and legacy oriented. With all this altruism oozing from our veins, we're stymied to take control of our land and farm business and start making radical adjustments. Successful farmers realize, deep down, that our intellectual and mechanical abilities are specifically endowed to us so that we may *participate* with our ecology like a masseuse, to caress the earth rather than conquer the earth.

Embracing the landscape is a distinction we have to make, or our fear and timidity will keep us from ever making adjustments toward an efficient working landscape. The reason I'm belaboring this point is because my discussions with frustrated new farmers quite often reveals this angst, this paranoia that our interaction with the landscape is ultimately arrogant and ugly. It's certainly not respectful. It certainly leaves a trace.

I must credit my farmer friend Michael Ableman with nailing this idea of participation. I've tweaked it a bit to *participatory environmentalism.* That is, as opposed to *abandonment environmentalism.* The differences between the two could not be more dramatic. As a farmer, I absolutely MUST embrace the former and disagree with the latter. To not understand this distinction way down in my innermost subconscious, is to doom my farming venture from the outset because it will relegate me forever to a tenuous, timid, paranoid interaction.

Okay, now that we've embraced participatory environmentalism, what does this look like? What does it require? If we're going to manage a piece of ground, what adjustments will we need to make?

1. Domicile. Where are we going to live? And if we live on the land, where is this habitation and what does it look like? I'll discuss this more fully in the chapter on living frugally. The main thing I want to point out here is that living on site has huge advantages and it's better to place something--even if it's a yurt--on the land than to live down the road somewhere.

A dwelling, at least if it's cheap and simple, does not require a large footprint and will not fundamentally alter the landscape. Of course, often a piece of land already has a house on it. Great. But if it doesn't, don't be afraid to build one. You'll never regret living on the land. And you'll never believe the inefficiency and cost of living off the land.

The idea of "going to work" is fairly modern. Historically, people lived where they worked. The cobbler lived over his shop. The baker lived over his shop. This is the way commercial districts were designed and served humanity very well until modern transportation allowed us to segregate the workplace from the home place. A farm is more than a workplace; it's a lifestyle. It's not a punch in, punch out kind of relationship.

Don't worry about destroying the pristine beauty of the place with a rudimentary shack. Farmers live in shacks all over the world. At least it'll be your shack. Beauty doesn't require landscape architects and buildings the average Home Owner's Association requires. Your simple abode will be a castle if you fill it with love and purpose. Don't be afraid to build a house on the land.

If you don't own the land, put up a portable house such as a recreational vehicle (RV), tiny house, mobile home, or one that's so cheap you can walk away from it and let it compost. I remember reading a *Mother Earth News* article back in the 1970s titled Kon Tipi about a family that built a $2,000 three-story house out of tall poles and a canvas covering. You know as well as I do that the kids in that family grew up with a can-do, adventuresome spirit. So it looks funny. Big deal. Even a shack can be tidy. A domicile, no matter how big or small, regardless of design, will not destroy the landscape. People living and loving on the land are the starting point for participatory healing.

2. Access. You're going to spend a lot of time and you're going to do a lot of things on this land. Whatever you can't access will be off the management table. You can only change what you can touch. If you can't get there, or if getting there is arduous, you simply won't invest in the adjustments. A good road system is what enables everything else to happen.

If your land already has old lanes or trails on it, chances are they're in a good location. Those old-timers knew a lot about efficient access. Following old deer paths or bison trails often yields the most appropriate lanes. Often you'll end up incorporating some of the old and adding some new. The secret to road longevity is to get the water off. Volume and velocity destroy roads. You can have volume with no velocity, or you can have velocity with no volume, but when you have both, you lose a road fast.

Broad-based dips (gentle valleys cut across the road to duct the water off) or water bars (gentle ridges, or dams, built across the road to duct the water off) are the normal way to move the water off the road. The steeper the grade, the more often the dips and bars need to be. Generally cutting the road into the side of a hill makes either of these construction techniques easier. Riding a ridge or going through a bottom (valley) work against being able to get the water off the road.

From a construction standpoint, these dips and dams should be angled rather than perpendicular to the road. At roughly 45 degrees, the water does not puddle up at the diversion. The reason this is important is because if the water stops flowing, it deposits silt and doesn't flush on out the edge of the road. Generally a puddle stays there and over time encourages a mud hole to develop as traffic slogs through the dampened spot.

The way to insure proper flushing is to come uphill as many feet as the road is wide to begin the excavation. If the road is 12 feet, come uphill 12 feet to begin. Of course, the excavated obstruction on the uphill side can be fairly low because not much water will be there. But as it moves to the lower side and the end of this hypotenuse, the dip and dam will need to be more pronounced.

I've seen extremely stable diversions like this that last for many years without any or certainly not much attention. If puddling occurs, either due to diversions being too perpendicular to the road or some other reason, placing fist-sized rocks on the upper side to create a permeable base will keep things stable. The big rocks allow water to move along the face of the diversion while also offering plenty of bed strength to support heavy traffic.

In gentle terrain, the road can double as a gutter to help fill ponds. The rule of thumb is 1 foot of drop per 100 feet of road keeps the water from going too fast to erode. At that grade, water flows about as fast as a gentle walk. To do this, of course, you'll need to in-slope the road to collect the water against the cut-in bank. Certainly the foremost design expert in the world on this concept is Darren Doherty, who founded ReGrarians, an Australian-based consulting enterprise. I highly recommend his work.

The whole object of this road gutter idea is to collect surface runoff and duct it into a pond. At strategic locations then, the in-sloped road shifts over to out-slope with a diagonal diversion into a pond. This works extremely well when you can ride the contour. Often, our farms are too small to include the necessary contour in order to duct water with continuity. Surveyors seldom consider topography. And generally, neighbors don't like you extending a road around their hill just so you can collect the water on the other side.

A balance definitely exists with this road gutter idea. It's certainly beautiful on a map, but one drawback is that it extends every farm trip. By the time you wind around all the contour road gutters, you've spent an extra hour and a gallon of gasoline getting someplace. So I think direct access is good and every time you can tap into the road gutter concept, the better. But don't complicate your time and maintenance budget unnecessarily with a cultish adherence to gutter roads. Strategic pond placement can still catch most of the water.

If a road must be steeper than this 1:100 ratio, I recommend out-sloping, which sends all the water to the lower side and gets it off the road. Obviously the road can't act like a gutter in this construct, but on steep grades it keeps the road from washing away.

I've never regretted a single road I've hired an excavator to cut into the landscape. Every one makes our farming more efficient and opens up more opportunities to interact with the land. One of the most revolutionary things we've ever done for our farm was in 1989 when we traded 30 acres of timber for 3 miles of excavated, all-weather road. Prior to that time, we had no access to 2/3rds of the farm. Once we had access, we could build ponds, cut timber, develop pig pastures, build picnic sites. It all started with access.

Make your roads wide enough for equipment and livestock movement. I think 16-20 feet is a minimum. Less than that and you'll inevitably be hitting fence posts along the edge as you transport portable shelters and farm equipment. A little extra width gives you some wiggle room for maintenance too. Productive land lost to a road is always more than compensated by efficiently being able to get from place to place.

Apply gravel to keep it passable. You can extend gravel by having your hauler place a block in the middle of his tailgate (before loading). When he starts to dump, the block keeps the middle of the load from falling out. By just placing the gravel in the heavily trafficked tire tracks, it will go a third farther. You don't need gravel in the middle of the road. Nobody is going to be passing at 30 miles an hour.

Perhaps one of the most strategically useful access developments is the pond/road combination in a valley. Let's say you have to build a lane across a valley. Orthodox excavation assumes culvert placement in the bottom and then covering with enough dirt to drive across. This is a single-purpose action.

What we do here at Polyface is to build a pond in the valley and put the culvert through the top. Rather than trying to drain the water away, we get a two-fer: pond and road. The top of the pond dam is the road. This often means you have to excavate more material than you would in the culvert-drain-crossing model, but it's not that much more and you get a pond in addition. Furthermore, pushing all the dirt out of the pond basin allows you to move both sides of the road higher on the valley sides. This means that rather than driving down into the bottom (or nearly) of the valley, the driving approaches are high because the road is built up on top of the dam. This completely changes drainage, erosion, and logistics, not to mention safety if you're pulling a heavy load. Moving access points higher on the landscape offers a multitude of benefits.

Whenever I drive on the interstate I think about this kind of technique. If it were used commonly in expressway construction, imagine all the ponds that would dot the landscape along the highways. If you're going to build up a road bed 20 feet, why not put the culvert at the top instead of the bottom and turn the causeway into a pond dam? With literally no more excavation than currently required, we could have thousands and thousands of ponds along our expressways, offering migratory waterfowl plenty of habitat, controlling floods, producing fish, and offering all the advantages of increased riparian environments.

Overall, excavation to create efficient access is some of the best money you'll ever spend developing your farm. A good operator with properly-sized machine can work miracles in transforming otherwise difficult-to-access areas into highly accessible spots.

Trust me, you'll never regret building a road on your farm. Whatever you have to move to put it in--a rock, a tree--don't worry. The shock of the initial change will give way to relief and joy when you can easily access that formerly hard-to-reach area. It's silly to pay taxes on land you can't even access.

3. Fences. Boundaries should be physical, strong, expensive fences. If you can't get everything secure right away, I recommend Premier's Intelli-rope as a temporary measure. It's a little pricey, but it's thick, strong, and highly visible. We've used it on rental properties up against heavily traveled public roads and never had a problem with an animal escaping. I still sleep better when the fence is physical rather than psychological, so I always recommend moving that direction, even if it's building just one roll a year.

Fences are the foundation of control. I suppose you could get by without fences if you didn't have any animals, but even produce farms often need protective fencing to exclude deer and rodents. The old adage "good fences make good neighbors" is certainly true. Just like roads, you'll never regret having good control infrastructure on your farm.

Internal fencing can all be electric. If you're running cows, one strand is plenty. Sheep, at least two strands; goats, at least three strands. One of the reasons I like all electric fence internally is because it's wildlife friendly. I hate it whenever I find a deer carcass hanging from the boundary woven wire fence because the deer got a foot caught in the upper square. But I've never ever seen a deer tangled up in an electric fence.

Your roads will become your arteries. Vehicular traffic, livestock traffic, and wildlife traffic all use the roads. Everything likes easy access. On the outside of the road, install electric fence. Put in plenty of gates, or access points. More is better than fewer; the more gates you have, the more flexibility you'll have. The idea is that one person can move animals from place to place by accessing the lane. Once the animals are in that lane, you can either lead them or push them anywhere, even from one end of the farm to the other. Without a lane, you'll dread a long move and so will the cows. If you think you're going to chase animals all over creation to get them from field to field, think again. You'll lose your religion and sanity. Build a road. Build a fence along it. Now you have control and security.

Fence out your ponds and creeks. This is not only good ecology, it's good public relations for your neighbors. Before too long, I imagine the government will mandate it, so go ahead and get it done. After that, run your fence along break points between ridge and hill, hill and valley, following the terrain in order to reduce all fields to no more than 200 yards across. If you're unsure about where to place a fence, just put in a temporary installation. The rule of thumb is that whatever you don't move in 3 years should be converted to permanent. Let function drive form.

Placement is extremely important because it sets up daily efficiency. If livestock won't flow through an opening, you've got problems. If timid animals ball up in a corner during a move, that's a problem. If the placement encourages wear and tear on a spot, that's a problem. Install only as much permanent fence as you absolutely need in order to make your portables work well. Over time, you can tweak these as your confidence grows.

4. Water. No water, no life. This is generally considered the number one limiting factor on farms. If you limit yourself to whatever rainfall naturally occurs on site in the abandonment mentality, the land will never produce what it could otherwise.

Fortunately, today we can run streams uphill through something called polyethylene pipe. It's one of the biggest breakthroughs in history. Obviously you can't get water in a pipe without a source. That can be a well, creek, spring, or pond. I'm a pond lover and have written extensively about ponds in other books, so I won't belabor it here. Perhaps few things are as disruptive and transformational in a landscape as a pond. Talk about drastic change, this is a big one. But few things can possibly compare with a pond for bringing life back where it's been struggling.

Permaculture, a landscape design system developed by Bill Mollison and Dave Holmgren in the late 1960s and early 1970s, is still the ultimate in this hydration scheme. Ponds need to be located as high on the landscape as possible in order to use gravity. On our farm, we've built a series of these over the years. Hooked together, they flow into a 7-mile grid of pipe that gravity feeds the entire farm. This requires no pumps and no electricity. When gravity quits, I'm out of here.

As long as the down-grade pipe does not rise above the

intake, the water will flow, even uphill due to the pressure from above. Few things excite me as much as watching water flow out of a pipe, whether pushed by gravity or a pump. It's as close to magic as anything I can imagine. Using petroleum to excavate ponds in order to impound floodwaters and irrigate in dry times gives more return on investment than anything else I can imagine, except maybe a chainsaw. Other than that, ponds have it hands down.

If it's illegal, build it anyway. Call it a diversity trough. Make a bunch of small ones instead of a few big ones. Call them terrain savers. We call ours fertilizer spreaders. "Yes, Your Honor, that's a fertilizer spreader. It sure makes the grass grow. Why is it that if I go out and buy a mechanical fertilizer spreader and bags of petroleum-based, earthworm-killing chemicals I don't need any permits at all, but if I build a frog-encouraging, grass-stimulating, neighbor-friendly pond as a fertilizer spreader, you're demanding a permit? This is as good a fertilizer spreader --in fact, better--than that one in the picture catalogue. It's not a pond--why, how you talk. This is the newest-fangled fertilizer spreader ever invented. Just because you haven't seen one and don't know the lingo doesn't mean it's not a fertilizer spreader. Are we using the same language?"

Today, every time we accumulate some extra cash, we build another pond. When you consider that prior to Europeans coming to America up to 8 percent of the entire landscape was covered in beaver ponds, you begin to realize just how much water we've lost. Farmers are our modern beavers. Whereas beavers always built in streams and creeks, today we can build higher on the landscape in valleys that only run when the commons is full and surface runoff occurs.

Limiting our ponds to these high-terrain valleys ensures that our impoundments can only add to the commons, not detract. By definition, surface runoff means the commons is full--the cup runs over--and holding that water uphill increases the commons rather than depleting it, which is what happens when you irrigate from an aquifer or river. I like things that everyone can do without depleting the commons. Let's bring back the water.

5. Buildings. Few things change a landscape as rapidly as buildings. But the alternative is far worse--rusty equipment and constant frustration over clutter. Buildings help organize the farm. Every farm will have tools, machinery, and inventory.

You can't store bags of mineral outside. Feedstocks rot outside. Everyone needs storage for stuff. Don't apologize for building sheds and outbuildings. They are the stuff of iconic farmsteads. What normal farm shopper doesn't love it when the ad says "and assorted outbuildings?" That's just the place you want to see. Of course it alters the landscape, but buildings protect the landscape from stuff sitting around looking like Fred Sanford's back yard.

This includes a farm shop. Farmers have to fix things. We need a place for the welder, torch, table saw, hand tools, work benches, a vice. These are such part and parcel of a farm that it seems silly to write about it, but I've encountered people afraid to erect a structure lest it desecrate the landscape. Hogwash. A shop, well outfitted and well worn is a tribute to the caretaking of the place as surely as a well-pruned vineyard.

I like to see all equipment under roof. It sure lasts a lot longer that way, and it adds order to the farmstead. Stuff just parked haphazardly out in the open creates a jolting first impression. Buildings don't have to be expensive or chic. Functional is fine. Build from rough sawn materials and straight-enough poles scavenged from the forest. But make no mistake, storage and protective buildings add hominess to the farmscape.

In your farmstead plan, be sure to include edible landscaping to soften the edges of the buildings. By placing some fruit trees strategically, planting some espaliered fruits, putting in a trellis, you can green up an otherwise stark building edge. Terraces on the down side of buildings located on a slope can completely change appearances. In my opinion, you can never have enough buildings. They always fill up. But keep them neat and tidy, painted and otherwise maintained, properly softened with trees, shrubs, vines, and garden areas, and these buildings will nest perfectly into the land.

6. Vegetation cutting. This includes everything from pasture mowing to harvesting trees. If neglect has created fields suffering from brushy, weedy, or sapling encroachment, a heavy mowing with a bush hog can arrest that degeneration and help you establish productivity. The way it happens to look the first morning you set foot on the land is not sacred.

If a tree is crooked, diseased, or mature, harvest it without remorse. That's our job. Doing so upgrades the woodlot and allows better specimens to convert solar energy into biomass. Part of our stewardship mandate is to keep precious sunlight from being wasted on diseased and junky biomass. This is not hubris, but rather quiet humility as caretakers who embrace a stewardship job description.

If something is in the wrong place, move it. If a tree is getting ready to fall on the fence or on the road, cut it down. Better to deal with it now than have it fall on you or a loved one or a cow. As protector, one of our jobs is to secure the landscape for long-term and continued functionality. A weedy woodlot is not nearly as productive as one growing thinned (weeded), highly productive specimens. Sometimes you'll find that the reason a field edge is returning to forest is because it's too hard to maintain as a field. Perhaps it's too steep or wet or rocky. Let that go. It's fine.

Don't be afraid to tackle an obvious unstewarded spot just because it's become that way and looks interesting. If you don't have something better in mind, leave it. But if you know you can use it to better advantage, change it. This is not pillaging; it's participatory environmentalism. So yes, go ahead and cut the tree. The others will thank you for it.

There you have it, folks, the adjusted landscape, the first thread for a successful farm business. I could write several how-to chapters on each of these adjustments, but I've written about that extensively in my other books. So I apologize if I've disappointed newcomers with my fairly perfunctory descriptions and explanations. But the point of this chapter--this thread--is not to go into the minutia of how-to; but rather to explain why a participatory attitude is essential for the successful farm.

As farmers, we must wrestle with these adjustments. I hope nobody comes away from this section thinking that I love to wreak havoc on every landscape. Far from it. I come meditatively, thoughtfully, respectfully to these adjustments. But a farmer who can't get beyond the necessity of adjustments, who thinks that these major landscapes adjustments must be wrong, will never be successful. That paralyzed farmer will fail as surely as a teacher who hesitates to instruct a student.

Realize that your arrival on your land is itself a big change. My presence on my farm changes everything. If I weren't here, it wouldn't be here. And if it weren't here, it would revert to its previous static state. That was a state that did not store floodwaters for slow release during the droughty summer. A state where almost every summer the soil cracked and dried, where the plants withered and sunbeams fell to the earth unused, unleveraged into carbonaceous biomass. It was a place where good trees were crowded by unthrifty, crooked, disease-prone weak trees.

These hands that can harm can also heal. Directing them with intelligent, deliberate, humble intellect is a win-win combination for the earth and for us. Creating working landscapes is the dream and hope of successful farmers. Now go hug your land; embrace your landscape changes, and be successful.

Chapter 2

Eclectic Awareness

Military trainers inculcate what they call "situation awareness" in new soldiers. It's a phrase that describes your surroundings, your context. Soldiers must know where they are, entry points, exit points, and resources. Anyone who ever watched a James Bond movie can only yearn for Bond's uncanny ability to know his vulnerabilities and strengths in any given situation.

I like the word eclectic rather than situation because it's broader and speaks to the process of acquiring situation awareness: an extremely wide-ranging understanding of context and applying this knowledge in real time.

Tai Lopez, my first formal apprentice and founder of The Knowledge Society, stresses during his entrepreneur seminars that the most common element among all successful people, across all vocations, across all ethnicities and demographics, across all cultures, is the simple act of reading. Put simply, successful people read.

But beyond that, they read widely, and that's where the eclecticism comes in. My mentor Allan Nation, who launched the grass farming movement with *The Stockman Grass Farmer* magazine, read a book nearly every day. But that's not all. What made him so interesting was that he read widely. Business, history, and his first love--railroad books. I've always been an avid reader with a constant pile of "to read" books stashed by my desk, and a stream of "just read" books exiting to bookshelves around the house.

These days, half of my reading is manuscripts to write forewords and I enjoy that because it's like getting a sneak peak under the hood of a new model before the car is available on the showroom floor. But beyond that, I read religion, relationship, leadership, business, politics, self-help, history, and of course food and farming material. Unfortunately, most people confine their reading to their profession or a narrow interest.

Most farmers in America only read trade journals--the ones featuring full page ads from industrial agri-business outfits. These present an incredibly jaundiced view of the agricultural situation. I'm always amused when some orthodox farming adherent accuses me of being ill-informed and narrow-minded, assuming that I could only come to my conclusions because I'm not reading Monsanto and Tyson material.

Goodness, if you're reading anything in farming, you can't help but read the orthodox material. It's on the front page of every newspaper, touted by every USDA bulletin, presented as gospel at USDA seminars, and preached at every land grant university. You'd have to live in a cave not to be aware and informed about the orthodoxy du jour. Flip on any TV news show, from Fox to CNN, and the newscast will laud the conventional line. Anyone with a modicum of public interaction is well bathed in orthodoxy.

The fact that I have chosen a different path indicates, inherently, a more eclectic approach. You have to seek out the alternative view. So when these orthodox apologists accuse me of being ill-read and ill-informed about their position, I like to ask them: "Have you heard of the magazines *Stockman Grass Farmer* or *ACRES USA*? How about the Weston A. Price Foundation's *Nourishing Traditions*? Oh, books, maybe you read books. Okay, how about Sir Albert Howard's *An Agricultural Testament*? Or perhaps Rodale's *Complete Book of*

Composting? Never heard of them? Where have you been, under a rock?"

I well remember doing a seminar in northern California around 2010 for about a hundred food and health inspectors. One of them had a dying wife who cured herself with raw milk, which stimulated a search for real food. This led her husband, a food inspector, to discover local farmers and the hurdles that government regulations posed to access integrity food. As a senior bureaucrat, this intrepid hero used his influence to convene a seminar of cohorts to acquaint them with the practices and philosophy of the integrity food movement. I was the designated presenter for the integrity food movement.

Knowing I was going into a hornet's nest, I asked for a show of hands: "How many of you have ever heard of the *New York Times* bestseller, *Omnivore's Dilemma*?" Not one hand among the hundred assembled food inspectors went up. I was shocked. I decided to pursue a little deeper, thinking maybe these weren't readers, but video watchers. "How about the acclaimed documentary *Food Inc.*? Anybody heard of it?" Again, not a hand went up.

Goodness, this was California, not Hicksville, Montana. Not a single one of those government officials had read, seen, or even heard of these two iconic masterpieces of my tribe. And so I used tribal metaphors to begin my presentation: "Okay, fair enough; you don't know anything about me. I'm a native American; you are the U.S. Cavalry. Before we kill each other, please come into my tent and let me tell you about me. And thank you for giving me this opportunity to powow with you." Mona Lisa smiles greeted me. They weren't sure how to take this farm boy debater. What fun.

We're all prone to in-grown thinking, to cultishness. How many Baptists read Presbyterian literature? How many Roman Catholics read Martin Luther's charges? How many Pentecostals read Reformed theology? We have to push ourselves to read widely, to read broadly. This happens in almost every group. The permaculture folks read their material; the biodynamics folks immerse themselves in that view; the industrial organics folks bathe in Wall-Streetified

Eclectic Awareness

applications and develop ultra-pasteurized organic chocolate milk. Talk about a jumble of misfitting words.

Exposure to other thinking brings either heightened conviction or a healthy questioning of our thoughts or deeds. One of the reasons I refuse to vote for either Republicans or Democrats is that they agree on one thing: shutting the minority parties out of public debate. What could possibly be wrong with letting the Libertarians, Greens, Socialists, and Constitutionalists join the forum? If these minority folks are as whacko as the mainliners say, then let the voters hear their foolishness and forever abandon them. But if these minorities have some excellent ideas, why not give people exposure to an alternative view? Only petulant paranoid pipsqueaks would deny a voice to these minority parties.

The truth is that we like boxes. For others and for ourselves. Boxes provide safety. At least we know where we are. To not have a box is to be vulnerable, but it's also the mark of a seeker. Many years ago I took on the moniker Christian libertarian environmentalist capitalist lunatic farmer in order to proclaim loud and clear: "Don't put me in a box!" I realize that if everyone became as rabid a non-joiner as I, we'd have far fewer functioning organizations. Perhaps that would be a good thing.

On my desk, the *Pork Checkoff* magazine lies right next to *Mother Earth News.* That is not a sign of wishy-washiness; it's a sign of eclectic awareness. Thomas Jefferson's vision of a yeoman proletariat, what he called the "intellectual agrarian," lives in my soul. The fact that our culture condescends toward farming as an occupation of the lower class, as something demeaning and appropriate only for D students and inarticulate, dull rednecks, is, frankly, embarrassing and painful.

This disrespect for the oldest and noblest vocation on earth has taken a toll in rural America. The cultural snobbery and prejudice toward farmers did indeed create what many call "rural brain drain." The best and brightest went to cities to seek their fortune and respect while the academically-challenged stayed home to tend the farm. Smarts is not everything, for sure, and certainly does not equate with common sense. I'm the first to acknowledge that academic prowess and stupid public policy are synergistic.

But we're not talking about public policy. We're talking about how you run your business, how you position your farm. That requires awareness of all the societal forces and trends coming to bear on your context. The fact that in the next 15 years more than 50 percent of America's agricultural equity--land, buildings, equipment--is going to change hands is testament to the fact that most farmers have been asleep to their context. If a farmer has no succession plan in his early 40s, for example, he's living in la-la land.

At the same time, wanna-be farmers' awareness of this huge agricultural exit presents an unprecedented and enormous opportunity. With the average American farmer now 60 years old and the demographic growing a few months a year, lots of land is opening up. The downside is that land prices relative to productivity are still out of whack. We'll talk about that far more in subsequent chapters, but for now it's important to understand that the only way you know about these trends is to read about them.

A farmer standing in a field is not alone, actually. Imagine a farmer in a pasture surrounded by 50 fat Hereford cows and their equally adorable calves. A more placid, beautiful scene can scarcely be imagined. But if you could see the threads of influence, like hooks, streaming into the picture it might change this idea of the independent farmer.

The pressures vying for influence are daunting. Somewhere at the county government center a group huddles around a table studying a map of the comprehensive plan. Arbitrarily, with the stroke of a pen, and often under the influence of developers, realtors, concrete companies and other interests, including rabid open-land advocates, these folks designate some areas as agriculture, some as residential, some as light industry. With the stroke of a pen, they can define what can and cannot be done in each of these arbitrary designations. In a moment, they can make someone a millionaire and a neighbor a pauper.

From one day to the next, this farmer's options change. If he's close to an urban area, tax pressure, public utilities, and heightened land prices suddenly change his situation. If he's in the hinterlands, the open space advocates get their pound of flesh by denying the farmer value added opportunities like a woodworking shop, eatery, or special events destination.

In another room at the county government center, the economic development administrators meet with industrial clients looking for special tax concessions and sweetheart employment deals to relocate businesses to the county. The economic developers peer across the farm landscape like predatory vultures, looking for vulnerable properties to rezone or acquire for industrial parks. In the room, all agree that farmland is an ugly stepsister, an economic backwardness on the county's progressive possibilities.

Meanwhile, in the commercial district, industry salesmen scheme marketing strategies on how to get the farmer hooked on their products. "Add pounds to your weaned calves," proclaims one marketing plan. Another promises an herbicide to rid the pasture of weeds. A fertilizer offer promises a 150 percent return on the investment. Assaulted by a phalanx of purchasing rewards, these businesses vie for the farmer's time and money. How will he know what to buy, what not to buy? Who's true blue and who's a charlatan? Some 85 percent of all decisions farmers make are based on the advice of a salesman. It takes a sharp cookie to sift through the chaff and make a good decision.

Every community is full of well-heeled folks, in-the-know, ready to take advantage of newbies. As Allan Nation always quipped, "if you go to a sale and don't know within 5 minutes who the sucker is, it's you." These let-me-help-you-folks lurk around the sale barn, the feed store, the grain elevator. They're ready to promise a deal to the unsuspecting. You need the wisdom of Job and shrewdness of Shylock to discern the right path.

Our farmer, surrounded by these interests, stands with his herd of Hereford cows, contemplating a new set of hooks. These are from radical environmentalists who lobby Congress for special taxes on cows in order to reduce atmospheric methane. Or perhaps today's issue is additional protections for elk who eat the farmer's pasture. Or wolves who eat baby calves. Or maybe the regulation du jour will lock up the nicest bottom ground by designating it a wetland because it floods once a century. As you look at all of those hooks dangling around the farmer in his field, you suddenly begin to realize the plethora of interests, like a giant tsunami, mounting up to take, define, confine, speculate, and tyrannize him, his family, his farm, and his livelihood.

As if that's not enough, the intern who just arrived to learn these pastoral farming methods opens up a Pandora's box of vulnerabilities. The farmer wonders: "Did I fill out the workmen's comp forms correctly? Will the county find out about that little cabin I built to accommodate the intern? What if the health department finds the composting toilet? Will the extension cords pass the workplace safety inspection? The old tractor with the once-a-day oil drip, will the EPA fine me and quarantine the farm as a hazardous waste site?"

These and a myriad other questions circle like a giant noose around the farmer as he stands with his herd of Hereford cows. How will he cope? What will he do? Read. Read. Read. Constitutionally guaranteed liberties mean nothing if you don't know you have them. Clever responses to all these distractions don't exist if you don't know about them. All the options and all the possible responses come as a result of awareness, of study, of reading. Survival depends on it.

If you're going to succeed, plan on reading often and widely. Doing so will immediately set you apart from the average farmer; indeed, the average American. Thomas Jefferson also said, "knowledge is power." Wisdom is the right use of knowledge and comes with experience. If you don't know, you can't make informed choices. Knowledge is the beginning of choice, which is, of course, the foundation for innovation. The lunatic fringe, Malcolm Gladwell's *Outliers*, always germinates the new ideas. Those new ideas always come as a result of seeking, which is marked by an avid interest in reading.

At Polyface we operate a formal intern and apprentice program, and I always know the ones who are readers. They catch on quicker; they have a greater interest in issues and subjects that come up for discussion. Debate and conversation is livelier when I'm interacting with a reader, and it warms my heart. It's not a guarantee of success, but if I were putting my money on a winner, I'd pick the reader over the non-reader every day.

Here are things I think you should read routinely.

1. The local daily newspaper. While often these are woefully inadequate for national and international news, they keep up-to-date on the biggest goings-on in the community. From the current brouhaha at the courthouse to the obituaries, marriages, and classifieds, it's your first peek into the neighborhood. I can imagine some millennials scoffing at this notion, viewing newspapers as obsolete. Okay, read it on-line. Digital is fine, but read the local newspaper.

Teresa and I have a refrigerator magnet in our kitchen that quips: "Journalism is the first draft of history." I admit to a terrible prejudice in this regard since during my junior and senior years in high school, then for a couple of years after college, I worked at our local newspaper, finishing as an investigative reporter before leaving full time for the farm September 24, 1982. I have a deep regard for the power of the press, the good--and bad--it can do for a society. But love it or hate it, it's the starting point for your community's information and collective thought.

You need to know what's going down, as they say. Remember, your farming enterprise is not an island. It's within a sea of interests, and knowing what those interests have in mind is paramount for success. Back in the 1990s I noticed a little obscure public hearing notice from the county planning commission. The proposed language in the zoning ordinance would prohibit sawmills in agriculturally zoned areas. At the time, I was dreaming about having a band sawmill.

I was the only member of the public who showed up at the hearing. When I explained that I had 450 acres of forest and would like to cut trees into lumber to sell to neighbors and use for building projects on my own farm, the board quickly appreciated an over-reach they had not contemplated. They were more than happy to amend the ordinance to allow on-farm sawmills for logs harvested from your own property. Had I not subscribed and read the local newspaper, that ordinance would have gone into effect and criminalized what we eventually succeeded in doing on our farm: buying a band sawmill and cutting our own trees into lumber.

Sadly, with many farmers beginning to die now, the obituaries can keep you abreast of changes in land administration. Perhaps a widow wants to lease out the land until the family decides what they're going to do. The local newspaper, for all its problems, keeps you abreast of local happenings. Attending events can give you hob-nob opportunities that can turn into positive business options. Knowing what's going on is a great conversation starter with your neighbors, a disarming technique to acquire the latest hush-hush information. Read the newspaper.

2. Tribal magazines. I could call this vocational or trade magazine, but I like the idea of tribalism. With their shorter articles than books, magazines offer a snapshot into your world. Obviously if you're into produce, read produce magazines. If you're into livestock, read livestock magazines.

These keep you abreast of current practices and interesting ideas within your specific field. New technologies and techniques come around every day. Obviously some magazines offer on-line perversions (I'm a Luddite), and that's fine. But look through these as they come in. Don't lay them aside and let them accumulate for a year. Study through them just like you would if you were being tested. You don't have to read every word; you can skip plenty. But look at every page so you don't miss something important. And don't leaf through it haphazardly while watching TV and listening to jazz in your earphones.

Read seriously, with full-on attention. I rip interesting articles out and file them in my 500-category filing system. You can read about this in *YOU CAN FARM*. No point in describing it again. The main thing is to have a way to index and retrieve interesting things. If you can't find it, it doesn't exist.

3. Books. Certainly start with books about your vocation, that speak specifically to the kind of farming you'd like to do. But that's just first base. Branch out into all the other areas I described earlier in this chapter. I think business books are extremely good for understanding trends, marketing, and team building.

Read about leadership and business structure. Read what the enemy says. Only then will you develop the depth and breadth necessary to be a professional in your vocation. People who read a lot are always interesting conversationalists or public speakers. Communicators lead their respective industries. If you want respect within your movement, read books: lots of them and many kinds.

4. The opposition. Yes, whether it's magazines, websites, or books, reading material that disagrees with your paradigm is good. It rounds you out and gives you common bridges to engage naysayers in conversation. This cuts both ways, of course. If you're an integrity food farmer, read *Saving the Planet with Pesticides and Plastic* by Dennis Avery. It'll do you good to get inside the head of an industrial food proponent. If you're a Monsanto devotee, read *The World According to Monsanto*. It'll do you good.

Much of the material our farm uses to inform our customers and define our marketing message comes straight out of industry publications. They incriminate themselves with great nuggets that we can use in our next communication with our customers. That's called original sourcing. Rather than quoting your friends about your enemies, reading the enemy's material ensures that your quoting and take-away is not distorted by someone else's negligence or bias.

People generally love to hear good analysis, to know what is right to think. If you position yourself as the fount of analysis, not only will you have an audience, but you will have loyal followers who trust you and will buy your farm products. The confusing language on labels and certifications is a tangled web of murkiness to the average person. If you can cut through that, and be a true north beacon for your customers, their loyalty will grow exponentially. Any time you can position yourself as the information filter and adjudicator your influence and marketability increase.

On purpose, I've spent a lot of time in this chapter promoting eclectic awareness through reading. I'm concerned that it's become tedious, like talking too long about a subject. But if I didn't think it was important, I wouldn't put the emphasis on it. Readers lead. It's that simple. People who read are interesting. They have more to contribute in any conversation; they have more and better advice, and they have a better chance of making good choices.

Now let's take a quick look at the other aspects of eclectic awareness. I won't belabor this as much. Successful farmers know their resource base: people, climate, topography, markets. Again, no farm is an island. If you're farming in Saskatchewan, you won't be growing coffee or pineapples . . . probably. With some of these new earth-sheltered greenhouses, though, who knows? You don't have to know about everything, but the more you know about your context the better off you'll be.

Garnering information is not a destination. You'll never finish learning. The day I don't learn something is the day I want to pass on.

Eclectic awareness includes knowing the different gurus of our movement. Anyone attempting to start a successful farm today should know certain names and certain bodies of material. If you're doing a vegetable operation, Eliot Coleman, Jean-Martin Fortier, and Singing Frogs Farm are iconic. I hate to mention names like this because inevitably I'll miss folks I admire and they'll feel overlooked. I apologize in advance and maybe I'm showing my own myopia in my list. To me, the quickest and most efficient way to become aware is to spend time listening, reading, and visiting the luminaries doing the things you want to do.

If you're in livestock, certainly Allan Savory, Jim Gerrish, Stan Parsons, Andre Voisin, and Allan Nation are high on the list. For general farming, J. I. Rodale, Ed Faulkner, Sir Albert Howard, Louis Bromfield, George Henderson, and Charles Walters come to mind. And for cultural anchoring, how about Wendell Berry, Barbara Kingsolver, Michael Ableman and Fred Kirschenman, Marion Nestle, Joan Gussow, Michael Pollan, and Gary Zimmer. In the culinary world, it's Alice Waters and Dan Barber. In the crop world, it's Colin Seis and Gabe Brown.

Once you get through all that, then you need to be up on nutrition and the science of food. Sally Fallon, Robb Wolf, Diane Rodgers and Natasha Campbell-McBride are making us aware of a whole new connection between food and health. Jo Robinson, Wes Jackson, Gary Paul Nabhan and Bill McKibben have carved out legacies of articulation and research that moves our movement's needle. And then you have the big isms, like Permaculture (Bill Mollison and Dave Holmgren as founders; additionally a host of great practitioner-teachers today), Holistic Management International, the Savory Institute, Regrarians (Darren Daugherty and Lisa Heenan), Elaine Ingham (Soil Food Web), and others.

These names and movements need to roll off your tongue like old acquaintances. A working knowledge of what they believe and what they've said will frame your paradigm. It's a wonderful thing to know something you're sure about. We live in a wishy-washy time where deep convictions are considered judgmental and narrow-minded. I heard a guy say once that if you don't believe something, you'll fall for anything. Don't be a milk toast. Hang onto something and be serious about it. Know what it is and what it isn't.

Know your resources, your terrain, your climate, your community. Circulate. Don't just hole up on your farm. Go visit people, attend conferences, and organize pot lucks. The mastermind concept is now practiced extensively in the business community and I think farmers need to use it. The idea behind it is that we should spend a certain percentage of time rubbing shoulders with people who are smarter than we are even in different areas of expertise. These folks challenge us to think things and do things we'd never attempt on our own. This is part of eclecticism. Stretching our dreams is as important as stretching our arms and legs.

On a day-to-day visceral basis, eclectic awareness means knowing where you are in the season, knowing what you have to do today, carrying a mental image of the tasks that lie ahead and organizing them so you move efficiently. What's the weatherman calling for? Go with the flow. If the weather prohibits doing this today, how does it fit in later in the week when conducive weather shapes up? We farmers hold a lot of loose threads every day. Wrapping them into a neat ball means understanding which ones are most important, which ones are critical, which ones can wait, which ones are strong, and which ones are weak.

Although situational awareness is a military term, it applies to all vocations and places. How much information are you taking in? What do you see? We stress to our interns here at Polyface to always assume something is wrong, because that's the only time you'll ever see it. If you just nonchalantly lumber along in your own little world, you won't see the tipped over water trough, the arcing electric fence short, or the cow on the wrong side of the fence.

I think an interest in many things actually exercises our sensory faculties to be aware of more things. It increases our peripheral vision and makes otherwise insignificant things more meaningful. Tom Brown, who wrote *The Tracker* and other survival, hunting, and awareness type books, could spot a turned blade of grass in a field. If I may take a swipe at modern video games, I think these games are absolutely compromising our awareness faculties and our observational skills.

Situational awareness is not cultivated when we stick our noses in a murderous video game. The house could burn down around us and we'd be oblivious. If you want to be a farmer, park the video games somewhere unavailable. Better yet, destroy them. You'll need full representation and function from all your senses. Exercise them.

Farming is a broad business. It spans mechanics, marketing, accounting, agronomy, health, construction and this is just the beginning. Perhaps the biggest mistake people make when coming to farming, especially after they've already spent time in another career, is to assume that since farmers are hillbillies, anybody can do it. Trust me, whatever field you're in, it will take all that education and experience and more to pull off the farming venture. That doesn't mean it's impossible; it just means it will be harder than you can imagine. Being aware of that is probably the first big step in cultivating eclectic awareness for yourself.

Understanding that you'll need to know a lot more about a lot more things is key to success. Most vocations today narrow down to where we know more and more about less and less. We're a culture of specialists with experts in narrow fields. Successful farming isn't like that. To be sure, industrial farming orthodoxy does try to narrow farming down to single species and single knowledge bases. But that narrowness creates its own problems.

For example, if we're just supposed to raise pigs, who cares about the blow out at the manure lagoon? That's not pigs; it's poop. Two different things; don't bother me with trifles. If I'm a cherry orchardist, all I need to know about is cherries. I don't need to know about hydrology or infrared insect communication. I just grow cherries. I would suggest that many of the shortcomings and problems of modern industrial farming are precisely because farmers believe the lie that they can operate in a narrow discipline and not worry about the whole picture.

To be successful long-term, then, we must know a lot about a lot. We must know the interrelatedness of things, the themes of truly great thinkers. That's how we build eclectic awareness, which keeps us balanced, innovative, and stable.

Chapter 3

Live Frugally

A machine you don't own can never break. The less stuff you have, the less it costs to maintain. How many fights start over stuff? Fights are expensive. Decreasing stuff is a good thing economically, emotionally, and environmentally.

Consumer debt in America makes the spendthrift addiction of the government's debt look like pennies. Whenever someone starts complaining about the national debt, I point out that we have a government that looks just like its people. That our national government reflects the financial principles of the citizenry simply indicates a cultural obsession with spending.

In farming as in anything else, expenses tend to rise to meet income. Wealth generally creates a commensurate increase in purchases, or expenses. The bigger house, bigger swimming pool, fancier car, and designer clothes all follow escalating income.

For many years, our affluent society--or what we think is affluent--has been spending a little more than its income. But the way to wealth is to spend less than your income, and to do it over a long period of time. Teresa and I are both blessed to have been born into families who spent a couple of generations living below their means rather than slightly above their means.

Live Frugally

One of the secret keys to the success of Polyface Farm is that I married a gal more frugal than I. That's absolutely the truth. And she handles the checking account. Generally any couple has one partner who is a bit tighter financially than the other. While I'm a big believer that marriages need to include the finances--no individual slush funds--I think it's wise to let the one that grips money a bit tighter be the one designated as "first responder" on the checking account.

Arguments over money are the single biggest friction in all relationships. Sometimes I think money harmony is a bigger imperative than any other. How we spend money may be the greatest physical manifestation of our value system than anything else in our lives. It's the ultimate visceral expression of our beliefs.

Show me how a person spends his money and I can tell you what he thinks. This is why e-commerce is now dedicated, if not predicated, to sifting through our spending habits to create profiles for targeted advertising. The more sophisticated these data systems become, the more specific the targeting.

When self-worth derives from possessions, we're in trouble. Of course, the whole object of advertising is to make us dissatisfied with our circumstances. If we just buy that product or service, suddenly our lives will be better. Unless we buy, we're missing out on a life worth living. This "keeping up with the Joneses" creates a spending treadmill that never satisfies, and like an addiction, it certainly impoverishes.

The reason frugality gets a whole chapter in this book is because I've seen countless aspiring farmers, and farm couples, fail due to lifestyle and spending problems. It's such a big deal it needs plenty of explanation. You see, dear friend, I'm the world's biggest fan of farming. And I want more than anything for folks who want to be farmers to be successful and to actually make good money. But that means you have to want it. You have to want it more than anything else.

When I hear the stories of people who win gold medals in the Olympics, their training regimen, and the discipline is downright impressive. Navy Seals. Wow--look at their training dedication. Why are there so few? Because most people can't mentally or physically dedicate themselves to such a singular focus. We get sidetracked. It's too hard. It's too self-depriving.

And so this discussion is a come-to-Jesus explanation of what it takes to launch a successful venture. People love to talk about Bill Gates and all his money. But who, when complaining about his wealth, is willing to pull all-nighters, live on junk food, and hole up in bathroom-sized apartments for years? The chances are if you're reading this book, you're aware of our family's success in this farming business and want to duplicate it in your own life. Congratulations on the dream. Kudos for the vision. I'm your biggest cheerleader.

Now let me tell you about "me and mama," as they say down south. We'll start the story with my great grandfather, Happy Smith, in Indiana. He and my great-grandmother had a shotgun wedding when it was a big deal and fled to the frontier in the late 1800s. We have their cooking pot and the pistol they carried for sustenance and security for three years.

After three years and with the baby in tow, they returned to their kin in Indiana and began farming. Happy was clearing land on his farm, extracting tree stumps with dynamite, when a charge accidentally went off prematurely and killed him. He had one son (born during the exile) and four daughters, one of whom was my grandmother, Nellie. She married Frederick Salatin (my middle namesake), whose family had emigrated from Switzerland, and they tried to hold the farm together but lost it during the economic downturn of the depression.

He and my grandmother moved into Anderson, Indiana so he could work in an automobile factory, but his heart was in farming. He never truly farmed again, but he had a magnificent large garden. It was bounded on three sides with a T-top trellis of Concord grapes, which he sold locally. He specialized in all sorts of berries: strawberries, blackberries, raspberries. He sold honey, eggs, and produce in the community during his working life. The garden was his retreat and rejuvenation from the factory job. His love for

Live Frugally

gardening and farming rubbed off on one of his five children: my dad, William.

As a non-wealthy midwestern boy, my dad's chances of buying land and starting a farm were slim--even in the 1940s. After an honorable discharge from the Navy post World War II, my dad got his degree in economics from Indiana University and met my mom there. She was in graduate school for Health and Physical Education.

My mom grew up in severe poverty during the depression. Her dad was an alcoholic who eventually abandoned her mom and sister. The three eventually went to live with relatives in a little duplex on the outskirts of Fort Worth, Texas: six children, three beds, one room. My mom put herself through college and then went on to graduate school.

Dad, always enamored of a new thing, dreamed of farming in a developing country and picked Venezuela. Mom, always ready for a new adventure, thought that sounded exciting. Dad learned Spanish in just six months at Middlebury in Vermont, hitch-hiked from Vermont to Mexico and spent six months honing his Spanish skills, then sat for the foreign civil service exam and passed on the first attempt.

As a bilingual accountant he was hired by Texas Oil Company to work with wildcat ventures in Venezuela. In those days, American companies would not allow American employees to marry until after they'd been on the foreign assignment for two years. After two years, Dad returned stateside and he and Mom married. They went back to Venezuela, built a house, and then bought a 1,000 acre upland tropical farm. They came stateside to have their two sons, my older brother and me, returning with supplies and American citizen children as expats, but tentmakers, living on the land. In 1959 a junta created political chaos in the country and as American capitalist pigs, our family was a prime target for the revolutionaries.

I'm leaving out a lot--a lot--of details, but suffice it to say that we were run off the farm and it was expropriated and we lost everything. Dad and Mom had put all their savings, everything, into that place and lost every dime. After attempting for six months to get protection, it became obvious that things were hopeless so we boarded a merchant freighter and landed back in Philadelphia on Easter morning, 1961. I was four; my brother was seven. Our sister was in the oven.

We found the cheapest, most worn out farm in the mid-Atlantic region, with a habitable house. Mom was pregnant and could not countenance the notion of an extensive remodel with all we'd been through. You need to understand that we landed on this worn out, gullied rock pile with not a single acquaintance in the area, no money, no equipment. The Ku Klux Klan burned some hay in our lane shortly after our arrival to let us know that as foreigners (we spoke Spanish and had just come from Venezuela) we were unwelcome in the community. How's that for a welcome wagon?

Mom took a job teaching at the local new high school and Dad stayed home as farmer and Mr. Mom. To this day I still relish stewed tomatoes, one of his signature suppers. Within a year Dad realized the farm could not pay both the mortgage and a salary, so he took an off-farm accounting job. Every penny--and I mean every penny--went to the mortgage. When others were getting TVs, we never had one.

When others drove nice cars, we never drove a car less than 10 years old. We never replaced a car until the one we were driving died. By died, I mean throwing a rod through a cylinder. I don't mean it wouldn't start. That happened routinely. Dad kept two sets of tires: one to use and the other to put on when the annual inspection came due. He'd put on the tires that would pass, then drive the car home and put the bald ones back on.

When others were buying pick up trucks, Dad thought multi-purpose. So he bought a neighbor's 1957 four-door Plymouth, took all the doors off and all the seats out. He sat on an overturned bucket (this was in the 1960s, prior to mandatory seatbelt laws) and drove the doorless sedan to tax clients, bow-tie flapping in the breeze. On the way home, he'd pick up chickens, a couple of pigs, a calf. The open inside of a four-door sedan is a big compartment.

When we kids would ride with him we sat on a chicken crate with our feet down inside. When neighbors were buying snazzy new PTO-powered hay balers, we used an old Case baler powered by a two-cylinder air-cooled Wisconsin engine with hand-start. I still can't believe nobody was killed by that hand crank. We had some exciting times when it didn't slip off the cogs. When it stuck everyone got away and waited for it to fling off like a bullet.

I don't want to belabor this more than necessary, but I think it's important to explain my roots to all the people who come here today and see four-wheel drive tractors, trucks, a band sawmill and functional buildings, that it hasn't always been that way. The main point I want to make is that today Polyface Farm is what it is, and can be what it is, BECAUSE of our frugal roots. That is what drove us to be creative.

I don't think Dad was the environmentalist that I am; he came to it through economics. He realized that chemical fertilizer was like a drug addiction. It wasn't that he hated fertilizer as much as he knew it was a vicious economic treadmill. So he looked around for alternatives. Our first large purchase was a dump truck. He brought home corn cobs from the local elevator and would spread them on gullies and rocks under the light of the moon.

For a decade we operated with a 21-horsepower 1951 Ford 8N, the little iconic gray and red tractor. The previous farm owner had bought it new and the wooden shipping box provided pig housing for a few years. The tractor came with the farm, along with the old Case hay baler, a three-point hitch mounted cycle bar hay mower, and a side delivery rake converted from horse drawn to tractor. That was it. That was the sum and substance of the equipment. . Most of our early work was planting trees, trying to get gullies to stop eroding (putting branches in them) and developing a rotational grazing program.

Putting every penny into the mortgage, by 1970 the farm was paid off. We were debt free, keeping an extremely small herd of cows (about a dozen mamas), a couple of milk cows, and a garden. Fortunately, although I had not met her yet, my wife Teresa was growing up in a frugal home where gardening, canning, milking cows by hand, and butchering poultry and hogs were foundations of existence.

Teresa and I met in high school in the early 1970s, became sweethearts, and began planning a life together. I wanted to farm more than anything and together we conspired how to pull this off. Her family had what I call a "real farm"--John Deere tractors, real hay making equipment, bank barn and income.

She knew how to pinch pennies. When I was 16 and she was 15, she made me a suit--I mean jacket, pants, the works, from scratch. Now there's a keeper. And with three brothers and no sisters, she knew how to cook, which is another good thing. I'm the entrepreneur but she's the anchor. I'm the dreamer; she's the realist. But she loved me anyway. I think she still does.

We knew we wanted to farm and we wanted to farm here on my family's place. Her brothers had dibs on her family farm so it was natural for us to set our sights here. The land was paid for, but it did not generate much income and certainly didn't employ anyone. With a handful of calves each year, our place generated enough to pay the taxes and put fuel in the tractor, but that was about it.

How to start? How to launch? Our situation was not unlike many of you. We had access to land and we were not in debt. That was positive, but we had no house, no salary, no strategic income-generating scheme. Fortunately, I had a wonderful background for entrepreneurship and a fall-back connection for work in town at the local newspaper. During my junior and senior year of high school I worked one night a week at the local newspaper, enjoying the mentorship of two crusty editors who honed my writing gift into what it is today. The debate team honed my speaking skills.

Dad, as an accountant, also kept a sharp pencil and taught me financial recordkeeping using my laying hen flock as the object lesson. With my background at the Curb Market through my high school years--getting up every Saturday morning of the year at 4 a.m.--I honed my sales pitch and marketing skills.

While in college, neither Teresa nor I ever had a car. I typed papers for money, earning a substantial part of my tuition. Her family scraped and paid for her education, including her accumulated babysitting money and odd jobs. When I graduated from college, I needed a car. While all my buddies were taking out loans to buy a car, I found one from a neighbor--a 1965 Dodge Coronet, 3-speed on the column, for $50. This was 1979. Folks, that's cheap transportation.

In my family, the apple didn't fall far from the tree. That Dodge was not too dissimilar to Dad's 1957 doorless Plymouth. Except that I left the doors on. No self-respecting college graduate, especially a professional journalist (I'd taken a reporter's job at the local newspaper) would be caught dead driving around in a 14-year-old $50 car. But I really didn't care.

You see, this is why I say what we spend money on tells a lot about who we are. If we're embarrassed by our car, our house, our jewelry, our clothes, we do not have a money problem; we have an identity problem. Of the many things I'm thankful for from my parents, I think perhaps the single biggest one--aside from my faith--is complete independence from peer approval. Anytime we think we need to spend money to fit in or to be accepted or to look like something, we're showing a serious character weakness.

It's called peer dependency, and it's deadly. One of the biggest compliments I ever pay a person is to tell them, "I think you're happy in your own skin." Too many people are unhappy with themselves. Goodness, look at the plastic surgery business, breast enhancement, face lifts, whatever. Radiate for who you are. Your stuff doesn't make you who you are; your heart makes you who you are. Your character doesn't depend on looks or clothes.

My Dad always told me the story of one of his aunts who routinely prayed for those "poor rich people trying to hold onto all their money." Who are you when stripped of money, clothes, degrees, and stuff? We don't need things to make us who we are. And I'm forever grateful that I grew up in a home where self-worth and self-image, had nothing to do with possessions, fashion, and marketing trends. It was all about character, about who we were when nobody was watching.

With our August wedding date looming, Teresa and I needed a place to live. She was living at her parents' home working in town; I was living at my parents' home working in town. We knew we wanted to be here. My family's old 1790 log cabin farmhouse had an attic. It was mud-dobber nests and tar paper with no insulation in the roof. Being originally wood shingled, the pitch was steep and it looked like an A-frame from inside with a massive 8 foot X 8 foot loose-laid stone chimney running up through it. The whole space was about 16 ft. X 30 ft.

The bathroom was downstairs. My younger sister was still living at home at the time, although she was in college. So we all sat down and decided to create an apartment in the attic and my sister would use Mom and Dad's bathroom downstairs, leaving Teresa and I the attic and the upstairs bathroom plus my old bedroom, which also happened to contain my desk. My old bedroom became our office and Teresa's sewing room.

At this point, I can hear the groans, "moving in with in-laws kills marriages." Or "what kind of guy never leaves his house with his bride? What a loser." You know what I say? "I don't care!" Mom and Dad were committed to giving us our privacy--they knocked on the wall when they came upstairs. I kept the woodstove stoked so Mom and Dad never had to get up in the middle of cold nights to put wood in it. The point is, you create specific expectations, maintain honor and respect, and make it work.

Did we have any arguments? Yes. But nothing much. Neither family could afford a new house; we simply did not have the money so we made do with what we had. This is the lesson here: make it work. It's not how you start out that defines the race; it's how you finish. We didn't get bogged down in the what-ifs of ten years out, or twenty years out. The shared space was a solution and we didn't get ulcers worrying about long-term solutions.

It was a way in. A cheap way in. With both Teresa and I working off the farm I'm sure we could have gotten a loan to build a house on the property, but if we had, it would have increased our income requirements and kept us from being able to make the full-time farm leap as quickly as we did. And it would have deprived Mom and Dad, who were getting older, from having youthful energy to stoke the wood stove at night.

It would have made two yards to mow instead of one. It would have required additional electric bills, gravel for another lane. The single biggest expense in life is your living situation. I cannot emphasize enough the importance of keeping your housing costs low. Live in a yurt, a hut. Immerse yourself in articles about people who build houses for under $10,000. Who cares if your friends think you're nuts?

While they worry about how to pay their credit card bills, you'll be free enough to pursue your dreams. Staying out of debt is the single biggest contribution you can make to your farming venture. I'm not opposed to all debt; sometimes strategic borrowing is necessary and good. But in general, I have a great disdain for debt and only use it if it will launch me to a better place. Living cheaply, we were able to save half of what we earned off the farm. Within two years of marrying, we'd saved enough money to live on the farm, without outside income, for one year.

On September 24, 1982 I walked out of the newspaper office. Nobody except Mom and Dad thought I was making a good decision. No college friends. No fellow employees at the newspaper. I don't even think Teresa's Dad and Mom thought it was wise, but her Mom was always supportive so she wouldn't have said anything, and her Dad didn't say much about anything at all. Church friends thought we were crazy. Everyone thought we were crazy. Did I say something about peer dependency? When everyone thinks you're crazy, you're probably onto the real breakthrough. Remember that.

I was confident I was making the right decision, but I had a kind of 50/50 faith in being able to survive. I thought I'd probably have to go back to town for another stint working for someone else before making the final break. I had sold some beef direct to folks at work and friends at church. I didn't think for a minute that we'd be able to survive beyond the year, but I reasoned that if we ran through our savings in a year, I'd get so many projects done by being home full time that our total time away could be chopped from perhaps 15 years to 10, or 10 years to 5. We're all familiar with the magic of compound interest, and how starting early is the important thing. It's much better to put in a dollar when you're 10 years old than $100 when you're 20.

Labor is the same way. Unleashing your energy when you're young pays much bigger dividends than unleashing your full energy when you're older. A year of unbridled energy at 25 years old is worth two at 35 and three at 45. So I reasoned that my year, as a 25-year-old, would pay off big-time in the future. And I was right. Which is why I encourage people to get a year's nest egg saved and then make the jump. What's the worst that can happen? You may have to go back to work off the farm for a little while longer. But you'll get to enjoy all the progress made in that laser-focused year.

As soon as I walked out of that office, I was free, not just emotionally, but economically. Suddenly I didn't have to wear nice clothes. That virtually eliminated clothing costs. We could shop at thrift stores. Today, if you visit, you'll likely find me wearing a shirt with a name on it--but it won't be mine. It might say Juan or Harry or Mike. At the local uniform cleaner's, we found shirts for 25 cents apiece. That's cheap clothing.

How many people wear clothes monogrammed with someone else's name and place of employment on it? Right now, I'm wearing a shirt monogrammed with George's, the factory poultry integrator. Who cares? It's cheap! Nothing pleases me more than when the interns all go down there and come back wearing these cheap shirts with all sorts of names on them. It's become a badge of honor. What's wrong with winning the game of "I can live cheaper than you?" That's one of the funnest games out there. It beats the Kardashians all to thunder.

Teresa and I watched our nest egg diminish over that year, but we hung on. I began cutting firewood to sell; that gave us some cash flow in the winter. We began raising broilers the second year and

Live Frugally

that developed a loyal client base. I helped a friend build fence and another plant trees. Both of those jobs brought in about $700 apiece. When you're living on $300 a month, a couple of $700 jobs go a long way.

We lived in that attic apartment space for seven years. Without TV. Others called it an attic but we called it our penthouse. Our bedroom was also the kitchen. We didn't have enough room for a bedframe, so we just put the box springs on the floor and mattress on top. It didn't keep us from producing two children. With the third-floor window view, Teresa could keep her eye on me. This was long before cell phones.

Rather than wondering what kind of living situation will make you happy, how about turning it into a game to see how cheaply you can live? Most of us like some friendly competition. Well, enjoy the competition of living in the cheapest situation you can imagine. A yurt. A camper trailer. Oh, but I'm getting ahead of my story.

With Rachel, our daughter, starting to walk, Dad and Mom graciously decided to replace my grandmother's small mobile home outside the yard fence (Grandma had moved down from Ohio when I was 15 years old and lived out her remaining 10 years here on the farm--cherished memories). Dad and Mom replaced Grandma's mobile home with a bigger, snazzier one. Dad had gotten quite ill by that time so having everything on one level was better, as well as central heat. We were still heating the big farmhouse with two wood stoves.

Teresa and I took over the big farmhouse and Mom and Dad moved 50 feet next door. That way we used the same well, same septic, and virtually the same lane. We had a big garden and essentially lived off the land. We milked two Guernsey cows, canned hundreds of quarts of vegetables, froze sweet corn, and raised all of our own eggs and meat. And we had our own firewood. We literally lived on $300 per month.

The $50 car threw a rod after a couple of years so we sold it for scrap for $75. How about that? We replaced it with a two-door

Dodge Dart that we bought for $1,200. We choked on the price, but it was a good deal

Dad passed away in February 1988. We were not quite six years into our full-time farming experiment, but by then we were confident we would make it. We had a handful of loyal customers and were cash flowing positive. Our car died right about the time Teresa's grandmother hung up her keys from driving. We bought her Grandma's car for $550 and drove it until a customer sold us his Lincoln Town Car for $1. I'm not making this up. That Town Car had a trunk about the size of a pickup truck.

We hired neighbors to come and do things that required significant equipment like front end loader work. I did a lot of shoveling in those days. To jumpstart the fertility, I'd drive down to a nearby cattle buyer's holding barns and bring home manure bedding. They'd pile it outside and it would compost. It had so much spoiled hay in it that it made beautiful compost. They would load it for me with a skid steer and then I'd bring it home and shovel it into a manure spreader. Each year for several years I hand shoveled more than 100 tons of that compost on the fields. It wasn't too bad because I'd park the manure spreader up tight against the back of the truck so I didn't have to shovel uphill.

Did I say something about leveraging youthful energy? You see, if I had gone out and borrowed money to purchase a front end loader at that time, it would have used up all of our cash. Financially, it made much more sense to leverage my labor, which did not depreciate and did not require oil changes and parts, than to over-capitalize with machinery.

As a general rule, don't invest in depreciable infrastructure until there is no alternative. As long as you have an alternative, whether it's working later, working harder, renting, or borrowing, don't put your precious cash into infrastructure. If it is something that inherently generates income, that's fine. For example, building a pastured poultry chicken shelter for $300 in order to turn $4,500 worth of poultry in a season is an investment in income generation.

If you're cutting firewood, investing in a good chainsaw is not an expense as much as it's an enabling tool that will let you generate more income. But when you make these investments, make sure they are necessary and you do just what is necessary and no more. If a 30 horsepower tractor will do, don't get a 50. If a $2,000 truck will do, don't get a $10,000 one.

Let's take a typical example, like a post pounder. Let's say you need to set, on average, 60 fence posts a year. If it takes 15 minutes to dig a hole and tamp one in, that's four posts per hour. If you're not fully employed, or if you can't think of anything more valuable to do with your time, then your time is essentially of no value. If you're neglecting earning $20 an hour if you spend an hour setting fence posts, then that's another story. But most farmers are not fully employed doing things that actually earn money. We do a lot of things that are hard to quantify, like the value of chopping thistles or fixing the gutter on the shed.

Let's assume you can buy a post pounder for $3,000. Chances are that you'll need a second person to operate the post pounder efficiently, although it can put in the post in 4 minutes (getting situated and positioned will take a minute or two). But it takes two people to do this; now you're putting in a post every 8 minutes per person, compared to 15 minutes by hand.

If for the sake of discussion we assume our labor is worth $10 an hour, the by hand method costs us $2.50 in labor per post (4 per hour). The post pounder is half that, or $1.25, so we save $1.25 per post with the machine. We have not put in any value to the diesel fuel or depreciation on the tractor and machine. We're just letting that slide for now. The question is this: at $1.25 per post, how long will it take you to recoup $3,000 for the post driver? Let's see, that would be 2,400 posts.

I can hear the objections screaming back through these pages: "Well, my second person is a friend who costs me nothing." Okay, why wouldn't your friend be willing to get some exercise and help put them in by hand? Tom Sawyer got his buddies to pay him to white wash his fence. Surely your friends will pay you for the joy of putting in fence posts. Explain that you're trying to install posts as cheaply and simply as possible, without petroleum and heavy machinery. Bring on the arguments; I'm ready. At 2,400 posts, putting in 50 per year, that's 48 years.

All of us, if we know many farmers, know that this scenario plays out everywhere. Now, if you're putting in 1,000 posts per year, suddenly the whole exercise flips over and you'd be a fool to not invest the $3,000 in a post pounder. I know farmers who put in only 10 posts per year on average but vow they must have a post pounder to be efficient. No, they're just drunk on what other people have.

The best option, of course, is to cultivate friendships with people who have post pounders. Let them buy the machinery and come over to your place to operate it. Give them some eggs and a Thanksgiving turkey and they'll be happy as clams. They get to horse around on a tractor all day--what could be more fun than that? --and you get your posts pounded in for no cash expense.

I have a friend in Florida who buys $400 junkers to pull around his eggmobiles. He just buys an old car, leaves it hooked up to the eggmobile, starts the car up to move the eggmobile, and when the car dies (about every couple of years) he just buys another one. It's not like he's putting any mileage on them. These are cars folks are buying to tear up in the local Demolition Derby. Have one of those guys keep a look out for you. You don't have to sell many eggs to justify a $400 clunker.

As our kids grew up, we didn't do Little League, ballet, or movies. Goodness, I think I've been in a movie theater about 10 times in my whole life. Ask Teresa why, and she's got a quick answer, "Why would I want to drive into town to the theater when I can just buy the DVD for $10 when it comes out, stay home, and watch it in the comfort of my old clothes and sitting on my favorite sofa?" Now that, dear folks, is a truly frugal lady. And at home you don't have to find a place to park, fight crowds, and put up with weirdos in the recesses of the theater. Good grief, when you think about it, there's really not any decent reason to go to a theater to see anything--except that everyone else does it and it's expected and it's a sign that you're "with it." Guess what? I don't care!

Call me the least "with it" guy in the world. I'll cry all the way to the bank. These things add up, and especially when you're starting, substituting labor for equipment, and homemade entertainment for tickets will pay big dividends in the future. Much of wealth generation is about deferment--delaying today's gratification for tomorrow's gain.

If you want to live like folks in town, then go live in town. Farming is hard enough without your emotional and economic equity being siphoned off trying to keep up with folks who get pizza delivery and spend half their time hanging out. You don't have time to hang out. You've got trees to prune, seeds to start, customers to court, hay to make. If you've chosen a farming vocation, be happy in it. If you can't find happiness in your vocation, do something else.

Livestock management guru Bud Williams used to say, "If you have to go on vacation to get away, don't come back." His point was that you should find the deepest satisfaction in what you're doing. If you don't find that, then change what you're doing. And if what you're doing provides your deepest satisfaction, why do you have to get away from it? Either change your vocation or tweak your vocation to where it does provide that level of satisfaction.

Simple pleasures are better than complex purchased ones. If your children can't get as excited about the frogs harrumphing around the edge of the pond as going to a movie, something is wrong. Any teacher working with a school garden program will tell you that earthworms and sprouting tomatoes generate far more interest than just about anything imaginable. An incubator hatching chicks? My goodness, it doesn't get any better than that.

You must eat, drink, and sleep your farm. It's hard enough as it is. Divided interests are deadly. Stay focused. Stay frugal. The dividends will slowly add up and you'll look back in 20 years and realize that your success was built on a foundation of simple lifestyle, simple living, staying at home, and focusing on holding onto every penny until it screamed for release.

One of our biggest challenges here at Polyface now is instilling this attitude in our interns. Many come with a 40-hour work week mentality. I've got news for you, dear heart, farming is not a 40-hour work week deal. Some weeks, when the weather is disagreeable, we only work 20 hours a week. Others, when the sun shines and hay's in the field, we work 90 or even 100 hours. We don't do very many of those, but when it's necessary, it's necessary.

We can't always attach a dollar figure to every action. People want to know what you're going to pay them for this or that. Let me ask you a question. Do you get paid to clean the toilet in your house? Do you get paid to cook dinner? Do you get paid to wash the car? Mow the lawn?

Lots of maintenance and upkeep things on the farm are like this. Chopping thistles--what's that worth? Cleaning the multi-flora rose off the fence? Thinning a new stand of trees? Because many of these things don't put money in our pocket today we have to be more careful about things that take money out of our pocket today. The stuff we buy needs to be tied directly to an income-producing procedure. If it's not, defer.

Plenty of new farmers buy an All Terrain Vehicle (ATV) immediately. We farmed for three decades before we ever had an ATV. Even then I bought one only because we had too many eggs to carry in by hand. But we bought it to facilitate a specific need. For years I hand carried eggs in from the eggmobile even when it was half a mile away.

Let's do our exercise again, like we did for the post pounder. Let's assume an ATV costs $4,000 used and requires $1,000 a year in maintenance and fuel. If we're bringing in 10 dozen eggs a day, easy to carry in two buckets, what's the added expense per egg for the convenience of doing it on an ATV? We might save a little time going out, but coming back loaded, we have to go slow to avoid breakage. Not much time savings there.

One of the biggest misnomers in economics is the idea that time is money. At Ranching for Profit schools, Dave Pratt has wonderful illustrations to prove, once and for all, that only money is money. Time is not money. If I do a project in half the time but the allocated expenses of speed double the actual cost, obviously time is not money.

So in our example, if with depreciation and maintenance costs my ATV costs $2,000 per year, at 10 dozen eggs per day for 250 days (the out-in-the-pasture season), that's 2,500 dozen. That means our ATV, at $2,000 annual cost, has added nearly $1 per dozen to the eggs! That's not profit; that's financial suicide. So here at Polyface, we purchased our ATV at the point when we were bringing in 60 dozen eggs per day. That saved us three walking trips and the cash addition per dozen was only 17 cents. And of course that's assuming no labor tradeoff or other positive economic uses for the ATV. At that point, it would have been foolish NOT to invest in an ATV.

If Teresa had her way, we'd have a day or two each summer designated as "early days" day. For those days, we'd use no ATV and hand shoveling would replace the front end loader. I used to hand shovel manure spreader loads out of the barn. I'd do two in the morning, come in and do desk work for a couple of hours around lunch, then go out and shovel two more loads before evening chores. That's four loads a day. In five days, that's twenty loads. When you look at it that way, it's not bad at all. And it's great exercise.

I'm all about efficiency and owning the right tool for the job. But you have to wait to purchase things when the time saved can actually return higher direct income elsewhere. If the expenditure is only going to free you up to do hard-to-quantify things, you won't be able to generate the cash to cover the outlay. Time isn't worth much if all you do with time savings is something else that doesn't generate income.

As soon as you have a machine, you need to put hours on it in order to cover the maintenance and depreciation. As soon as you have a building, you need to use it, stuff it full, preferably with income-producing things. Use all the space. Put screens across the rafters and dry flowers or herbs up there in the hot peak of the shed. Part of being frugal is not putting up with wasted time or wasted space. Meter both out carefully and efficiently.

The farming community is full of people with expensive nonessentials. Yes, we have tractors. We have six of them in fact. But we generate $400,000 of income per tractor. I know one farmer who has eight tractors and only generates $10,000 of income. He doesn't rely on the farm for income and subsidizes it heavily from his off-farm job, but sheesh, that's crazy. How many farms have a tractor or two but only generate $100,000 in annual gross sales? Folks, if you're going to have a tractor, you need to put 500 hours a year on it and generate some income. Otherwise, borrow, rent, beg, or do something different.

The average farm in America requires $4 worth of depreciable capital (buildings and machinery) to turn $1 in annual gross sales. On our farm, we average 50 cents to $1. That's an 800 percent difference. That's not bragging; it's just stating a frugal fact. It means we squeeze more out of investments than the average. Our labor costs are higher and we'll talk about that in another chapter. But labor does not depreciate and is in real time today dollars. The money in labor is not spent up front as a capital expenditure. Labor generates cash flow, or at least that's the goal.

When Teresa and I began processing broilers, we had a simple set-up with an automatic scalder and automatic tub picker. I'd get up at 4 a.m., fill the scalder, turn it on, and go back to bed. We'd get up a few minutes before 5 a.m. and process 100 birds by 7 a.m. Then I'd go out to do chores while she went in and got the children up, breakfast, and put the house in order.

I'd come in at 9 a.m. with another 100 birds. I'd grab a quick breakfast while the scald water came up to temp and then we'd process the other 100 birds by noon. That gave us one hour for clean-up before the first customers arrived at 1 p.m. By 4 p.m. the last customer left and I'd go do chores and Teresa would fix supper. We ate at 6 and died at 8. We'd do that about three days every three weeks.

One of our big concerns today, with nicer equipment and more people, is that our interns never see what an efficient two-person crew can accomplish. Since they're used to an eight-person crew, they assume that you just can't process chickens unless you have eight people. Starting out, it's much better to just process with two people more days than hire help and try to cram it all in on one day.

When you're starting a business, the goal is to spend more time doing things that actually pay you money. Processing pays you money. No advantage exists to get it done faster if it doesn't enable you to do something else that earns money. Hiring people to get it done faster simply reduces your capacity to be fully employed at tasks that actually return income through real time.

The main takeaway from this chapter, I hope, is a profound understanding of saving for down times and not spending until it's an absolute necessity. That's the practice. In theory, it's all about justifying every expenditure on its own merits rather than needing something because your neighbors have it or the profession assumes it.

Thank you for indulging me this walk down memory lane. I hope this hasn't sounded like the T-shirt that says: "The older I get the better I was." I tried not to embellish things too much. Yes, Teresa and I had a lot of support and a great foundation on which to build. But plenty of people have started with family support and land only to lose it a decade or two later. For sure, we're both blessed because we grew up in frugal families. My family added a true rebel spirit to its frugality that didn't care a lick about what others said or thought. That is wonderfully life affirming and I recommend it highly. Now pinch those pennies and make them work for you.

Chapter 4

Can-Do Entrepreneurial Spirit

We're hard wired to be negative. I can never remember coming back from a trip to town and announcing enthusiastically: "I hit 5 go lights!" We call them stop lights because we remember what stops us far more than what makes us go.

We love the negative, rather than the positive. How much easier is it to believe negative gossip than to give the benefit of the doubt? We have the saying that it takes years to develop a reputation and only one mistake to destroy it.

I do a lot of conference speaking. At the end of a performance I can have a standing ovation but perhaps one person will come up and complain about something I said. Guess what I'll remember when I go to bed that night? Not the standing ovation, but the negative comment. It just sits there like a sore as I examine it from every angle, nurse it, massage it.

Do you know how much easier it is to say no than say yes? No is easy. Status quo is easy. What is, is easy. Change is fraught with difficulty, the unknown, risks. This is one of the singular helpful nuances of formal debate. I debated on the interscholastic team during high school and the intercollegiate team throughout college.

The two sides are called affirmative and negative.

In policy debating, the resolution is always stated as a change in the status quo. The affirmative, which is in favor of the resolution, must support change. The negative, of course, is opposed to the resolution, or the change. The negative argues in favor of the status quo, including how the status quo is the best or can fix whatever problem the affirmative alleges as a need to change.

The upshot of all this is that the negative enjoys presumption, which simply means that we presume what is already, is best. It's familiar; it's known. The affirmative has two options to overcome this presumption. One is to show that the status quo is producing enough harms that it's untenable. In debate parlance, we call this "dead bodies on the flow." A flow chart is how debaters take notes of the flow of the arguments throughout the debate.

The other way the affirmative can overcome presumption is by showing enough benefits from changing the status quo. The status quo may be okay, but if we can have all these benefits, it's worth making a change. In either case, whether it's harms or benefits, that basic critique of the status quo is called "building the case, " or the "case side" of the debate. In order to win, the affirmative must win the case; otherwise presumption says vote negative--stay with what we know.

But that's not all the affirmative must do. Said another way, the case can be called the complaint. In order to win the debate, the affirmative must win a second argument, and that's known as the plan, or the solution. I have always appreciated this about debate. You can't win by whining. You have to win both complaint and solution. The negative side, therefore, can lose the case (affirmative wins the complaint), but win the debate by impugning the solution.

Many ways exist to do this. The solution might not solve the complaints alleged by the affirmative. Perhaps the plan will be circumvented or simply be unworkable due to clumsiness or bureaucracy. Sometimes the negative can win simply by showing that the affirmative's plan will create worse consequences than the current poor status quo. These are known in debate parlance as disadvantages, or "DA's".

Without getting too bogged down in formal debate minutia, isn't it telling that formal debate theory is predicated on both complaint and solution? When applied to real life situations, these debate nuances come to life. For example, in a courtroom, the defendant is negative (innocent until proven guilty) and the prosecution is affirmative (change in status quo, from innocent to guilty) and bears the burden of proof (beyond a reasonable doubt). Why? Because the risk of changing status (innocence to guilt) is serious business. The charges (complaint) must overcome the presumption of innocence.

In marketing, the seller is affirmative (change in status quo, from not owning something to owning something). The buyer is negative. The seller must overcome both the reticence to change status, from non-owner to owner, and provide enough positives to a change in status that it puts the buyer beyond concern regarding something new.

I've done many debate workshops with teens and always point out that the rules are the status quo. That means if you want the rules changed (time of curfew, for example), you are affirmative and must show both the benefits of a change and a plan to overcome the presumption of the status quo. All the presumption lies with the parents who have created the status quo. The necessity for the affirmative to provide a workable, accountable, credible plan really creates some aha! moments for these young people as they see the visceral application of debate theory. Often they see themselves armed with super powers going into the next argument with their parents. But I also teach the formality of debate, that it is a respectful, ordered discourse, not a fit or tirade.

All this to say that anyone who starts a farm, markets products, or wants to do something different is affirmative, which is much harder to maintain than simply being negative. The burden of proof is always on the affirmative; all negative has to do is question, question, question, planting seeds of doubt. That's enough to win.

For many years now, in our culture, the status quo has marginalized farming emotionally, economically, and certainly environmentally. The orthodoxy is that bright-eyed, bushy-tailed entrepreneurs can go into any vocation except farming. And within the farming community, the orthodoxy is all about factory farming, industrialization, chemicals, GMOs and exports to far away places. The point is that the status quo, on many fronts, is quite well defined and that's the home of presumption. Being affirmative is fraught with proof, with burdens, with plans.

Who wants to be affirmative? It's more work! People with a can-do spirit want to be affirmative. Anyone familiar with business books knows about the Peter Drucker learning curve. In any enterprise, Drucker said you enter at a certain point, with a certain amount of information and expectation, capital and other resources, but then the curve goes down and not up. This valley, where our expectations change and we realize how ignorant we were, and when we use up our capital and friends abandon us, usually lasts 3-5 years.

That's a long time to stay with it. Virtually every success story has this valley in the early days. Financial get-out-of-debt guru Dave Ramsey constantly reminds his readers and listeners that he went bankrupt in his early days. My own experience includes some extremely hard and precarious times in the first years.

The opposite of success is not failure; it's quitting. As iconic radio broadcaster and optimist Paul Harvey used to say, winners "get up one more time than they fall down." Everyone stumbles, makes mistakes, and goes through valleys of despair. What separates the winners from the losers is that the winners hang in there through the valley until it turns upward. Most start-ups quit right before the curve turns upward at the end of that valley.

In the Drucker curve, eventually the valley turns upward, goes way up higher than the point of entry, and eventually rounds off again as success creates a new set of challenges and learning curves. During this valley we must surround ourselves with people who believe in us, with information that will teach us, with mentors who will challenge us. We need a tribe.

All of the support networks for fighting drug addiction, alcoholism, obesity, and abuse recognize the importance of a tribe. Sometimes it's family members but often it's not. I get calls from distraught parents: "My daughter applied to your internship program. Please explain to her that this is not a vocation worth pursuing. We didn't send her to college to get a math degree to go dig in the dirt." Obviously, Mom and Dad didn't read *YOU CAN FARM*, and won't be a part of her support network.

The prejudice against farming is palpable and it's up to you to surround yourself with a support group. I'm a firm believer in the old saying that when the student is ready, the teacher will appear. If you start seeking, you'll discover. What you can't afford is a victim mentality. Modern culture applauds, nurses, embraces, and worships victim-think.

Whenever I do a farming masterclass, regardless of where it is, I start with this context: "I'm from the Shenandoah Valley of Virginia, where our frost dates are May 15 and September 15, our rainfall is 31 inches per year, and on our 650 acre farm we have 1,000 feet of elevation difference. When our family came in 1961, we had 16 foot deep gullies coursing down the hillsides, leftover from two centuries of plowing in a winter freezing, summer baking, torrential rain storm hard agronomic environment.

"In fact, when Dad developed a portable electric fencing system in the early 1960s, we didn't have enough soil to hold up electric fence stakes. He poured concrete in used car tires and pushed two half-inch pipes into the concrete: one straight up and down and the other on a slight slant. My brother and I were just big enough to heave these off the back of the tractor platform as Dad drove slowly across the field. Then he'd go back and stick electric fence stakes in the appropriate half inch pipe so he could build electric fence.

"I remember as a child walking the entire farm and never setting foot on a piece of vegetation. It was known as the armpit of the community. We couldn't support 10 cows. Organic matter averaged 1 percent and water ran off muddy from the landscape in even a slight rain. Dad called in both private and government agricultural consultants to explain how to make a living on this farm. Their universal advice was to plant corn, plow, apply chemical fertilizers, borrow more money to build silos, install a feedlot, graze the forest.

"Today, our organic matter is more than 8 percent, even flood waters run clear, we graze 100 cows instead of 10, the gullies have stopped and it's arguably the most productive farm in the region. That's not bragging on us; it is bragging on the redemptive capacity of nature to respond to the touch of caress rather than a Conquistador."

That's the opening salvo. The truth is that whenever we hear someone talk about their success, we (including me) always assume that they had some sort of unfair advantage. We rationalize that this guy is successful because he had a better education, a better piece of land, neighbors who loved him, customers who couldn't wait to pay higher prices for good food, a spouse who loved whacky ideas, a better tractor (one that runs), a better bull, better rain . . . . The list never stops. Folks, this is whining; it's victimization. And it has no place in the lexicon of the can-do spirit.

People call me from all over the world for advice on how to do something. I start explaining and their first reaction is "well, that wouldn't work because . . . " So I offer another possibility, to which they reply, "well, that wouldn't work because . . ." After about the fourth round of this, I just have to say, "well, you're right, it won't work there." What I mean is that you won't ever be able to make it work because you've convinced yourself that it's too difficult or you're too special or your circumstances are too different.

I always tell audiences that if I did my presentation right here in Swoope, Virginia, the farmers here would fold their arms across their chests and grunt in unison, "well, that's fine for you, but it won't work here." All of us feel threatened by new ideas. Few things qualify as a new idea more profoundly than the notion that you can actually make a nice living on a small farm. That's not in the paradigm of the average person.

I hope nobody reading this interprets my encouraging words as a recipe for comfortable and easy success. Nothing could be further from the truth. It won't be easy. Nothing worth having comes easily. But something worth having is not what most people do. The current vocational psychologists agree that 80 percent of Americans hate their jobs.

What does it say about a culture as innovative as ours when only 20 percent of people truly enjoy their jobs? I'd say that provides plenty of room for innovative vocational adjustments. Generally the dissatisfaction is not monetary or even career choice. It normally centers around the workplace, recognition, autonomy, and attitude among co-workers. It's relational and emotional, not financial.

I've always said that our support group here at Polyface has always been our customers, not our farming community. Most of the farmers call us Typhoid Marys and bio-terrorists. For more on this, read my book *THE SHEER ECSTASY OF BEING A LUNATIC FARMER*. Early on, our family sought out some biological farming mentors. They aren't famous and aren't even names you'd recognize, but they encouraged us and provided a sounding board.

They did not think we were nuts. They believed in our ideas and moved us progressively in our thinking. I stand on their shoulders. But I had to seek them out. I had to go visit their farms. I had to invite them to dinner. I had to give them a Thanksgiving turkey. I had to buy some of their stuff. All of us have a protective shell around us to keep us from being scammed or investing time and energy in scofflaws. So you have to romance them into your life. You have to show yourself worthy in action and spirit.

Every time I think I've heard the most difficult success story, I hear another one that comes from worse circumstances, more impossible odds. The real success stories do not generally come from people born with a silver spoon, but from people who overcome all the odds to create a legacy. The stutterer who became a renowned public speaker. The destitute orphan who becomes CEO of a Fortune 500 company.

We've all heard about them and read about them. But we universally think those are someone else's story and can't be our own. Lots of times I think our own success here at Polyface is simply because we didn't give up. I could write a long chapter about the many folks who have started with a big splash and then gone out of business. In fact, those outnumber the ones who stuck with it through thick and thin and punched through to the 10-year mark.

Determination in a good thing, a noble thing, is the greatest asset a person can possess. Being the last guy standing often means you succeeded when others failed. Longevity can cover a multitude of sins and errors. When you're going through the fire, time seems frozen. But looking back on it, you realize what a blip in time and space that hardship or crisis really was. Those hardship experiences build your confidence and affirm that you actually can pull this thing off.

You can't Google experience. Adam Taggart and Chris Martenson's book *PEAK PROSPERITY* has a wonderful chapter titled *Knowledge Capital*. They point out that mastery requires repetition. They build on Malcolm Gladwell's *OUTLIERS* idea of the 10,000 hours and 10 years to become an expert.

Taggart and Martenson say this, "Developing experiential knowledge is all about appreciating the vast difference between understanding how something works in principle and how it works in actuality. You don't 'know' something until you've done it, and done it many times. . . A novice works under supervision, tackling increasingly complex tasks as skill level increases over the years, until the novice is himself a master. Developing such expertise takes time, effort and dedication. When acquiring a new skill, it's going to require a lot of repetition before you not only know how to do it well, but have encountered the unexpected edge cases and know how to handle them appropriately. It was for a good reason that apprenticeships during ancient times lasted seven years. They saw that it took that long to truly master a tradecraft."

While all of us intuitively know that this is truth, actually doing it is quite a different story. I always tell folks I can teach someone how to gut a chicken in about 10 minutes. But you won't become a master until you've done a young bird, old bird, rooster, stewing hen, full crop, broken intestine, full intestine, big, small, young, old, rainy day, hot day, cold day, with a cut finger, with a hurting back . . . You get the point. These are the nuances that separate novice from master. Unfortunately, no shortcut exists.

You just have to plug away at it. No victimhood. No excuses. And no entitlement. Let's pause on that one. Sometimes we farmers get drunk on our own importance. We proudly display bumper stickers that proclaim: "No Farms, No Food." Or "Have you hugged your farmer today?" Certainly farmers have been marginalized for a long time. I get that. But pendulums never stop at the center when correcting for swinging too far one way.

The overcorrection on farmers' self-worth I'm afraid has created a swing to the other side. Many farmers, especially those devoted to local, beyond organic, nutrient dense fare, carry a chip on their shoulder that says: "I'm saving the world and since I'm doing such a noble, righteous thing, you should buy my food, get the USDA to give me grant money, and subsidize me like you've subsidized all those bad guys all these years." I understand how those of us who have been kicked around by the bad guys want our place in the sun, a place to proclaim our righteous indignation and get some applause for a change.

I understand that feeling. But this kind of thinking can quickly cloud a can-do spirit and turn it into a whiney "you should love me because I'm special" kind of attitude. Too many folks exhibiting this mentality fritter away their time sitting in government offices and filling out grant forms rather than pulling on their boots and getting on with the work at hand. Government money always comes with strings attached, and it always comes so tightly constricted with dos and don'ts that generally you'd be better off just going out and doing what you want to do rather than cost-sharing with a bureaucrat.

I've never regretted refusing to participate in agriculture grant money. Easy money never makes you creative; hard times make you creative. I'm thankful Teresa and I didn't start with a bank account. We had to figure out how to do it cheaper and more efficiently. We couldn't afford to just pour money into what the experts said we should do--the experts who never operated a successful stand-alone farm, by the way. Hard times are not the problem. The problem is depression, frustration, and fear.

So what are the fears that keep us from doing? At the risk of drawing this chapter out too long, let me summarize what I consider to be the 7 deadly fears. Some of this will be expanded at other points in the book, so don't get frustrated if they aren't fleshed out enough for you. But I wanted this together in compressed form for your cogitation and planning.

The first fear is KNOWLEDGE. How do I do this? What do I do first? Where do I start? Here are solutions:

1. **Start with what you like.** I always tell folks to produce things you like to eat because you might have to eat your way through your inventory. Do what interests you, not what someone else says you should do. Follow your passion.

2. **Do the opposite of industrial anything.** Conventional advice always worries me. Advice from people who are not doing what you want to do should always be suspect. Remember the advice my dad received in the early 1960s--graze the woods, build silos, put in a feedlot, plant corn, build a confinement chicken house. It's all a bunch of hogwash.

3. **Start small.** Think embryos. Ask "How small can it be?" Not "How big can it be?" Innovate with prototypes. Test things before going whole hog.

4. **Start something . . . anything.** Fill your space if you live in a condominium. Grow patio plants; put a beehive on the roof. Movement creates movement. Don't worry about 10 years down the road. Do something today and tomorrow will take care of itself.

5. **Good enough is perfect.** Function is more important than form. Make it work and then worry about how to make it pretty.

6. **If it's worth doing, it's worth doing poorly first.** Imagine Thanksgiving dinner with the family. The newest addition, little Sally Ann, 8 months, is crawling around on the floor. While

the adults visit, Sally Ann scoots around, grasps a chair leg, and struggles to a teetering partial-standing position. Her mother sees her first and exclaims to all assembled: "Oh, look! Sally Ann is standing up! That's the first time!"

Sally Ann suddenly realizes she's the center of attention, smiles rapturously, then loses her grip and plops down on her bouncy diapered bottom. What do all the adults do? You know what they do. They crowd around, point their fingers, and with shrill prejudice hiss, "Sally Ann, if you can't stand any better than that, just quit!" NOT! Of course they don't. They quickly gather around, cooing encouragement to try again, helping Sally Ann find the chair leg and boost herself up.

The truth is we don't walk well, talk well, write well, poop well . . . oh, I guess we poop alright but we just don't know where to put it. We don't do anything well initially; we do it well after practice. And so grandma's haunting admonition "If it's worth doing, it's worth doing right!" is absolutely WRONG! The truth is that if it's worth doing, it's worth doing poorly first. So next time you're paralyzed by the fear of not doing something right, promise me you'll pull up your diaper, look in the mirror, and tell yourself: "If it's worth doing, it's worth doing poorly first." And then go do it.

The second big fear is ACQUIRING LAND. How do we afford land? Where are we going to farm? How do we get a toehold? Here are some solutions:

1. **Start with a portable farm.** Fortunately, soil can be healed relatively quickly. Think about portable infrastructure, including multi-use equipment and buildings. Customers are portable; the grain elevator is not.

2. **Equity is in management, information, and customers.** Most farms operate on large capitalization and depreciation schedules. Invest in people, management, customers, and skill.

3. **Piggyback by adding enterprises to existing farms.** Run pastured poultry on a grain farm or under an orchard or on a cattle operation. Tuck an orchard on a steep hill not currently being used for anything.

4. **Partner with those who have land.** A tremendous amount of farmland is becoming available for rent. Rent is based on productive value; purchase is based on non-farm market value. How about caretaking a place? Elderly couples often seek partners to caretake their place, sometimes in exchange for some meals and housecleaning. Public lands or trust-owned lands have been successful germination acres for farms.

The third big fear is FINANCES. What happens when I don't have a paycheck from my off-farm job? Can we get up and running fast enough? What if we run out of money? Here are some solutions:

1. **Be debt-free.** Live cheaply and cut living expenses. We already talked about this at length. Work off-farm an extra year if necessary in order to get out from under debt.

2. **Create a 1-year nest egg.** This takes off a lot of pressure. What's the worst case scenario? Go back to off-farm work for a year or two.

3. **Fill time with piece work.** Most start-ups are not fully employed. Are you handy at anything that could be compensated? Some light construction. Building fence. Painting. Planting trees. Doing honey-do projects for neighbors who spend all their time commuting to town. Run a band sawmill.

4. **Create cash flow.** Be careful about producing items with a long turn-around or a once-per-year harvest payday. Think multi-season, complementary enterprises, and consumables that people use up and have to buy over and over.

The fourth big fear is LABOR. Who will do the work? What if we want to go away for the weekend? What if I get sick? How can I afford employees? How can I get it all done with only 24 hours in a day? Here are some solutions:

1. **Love people.** I know this is not a normal attribute of farmers, most of whom are farmers because they don't like people. But get over it. You can't do it all yourself. All the gifts and talents necessary for a successful business do not grow on the same pair of legs.

2. **Incentivize the team.** Create commission-based partners, not employees. Use performance-based remuneration, not wages. Develop independent fiefdoms, including for your children. I've written about this extensively in my book *FIELDS OF FARMERS*.

3. **Build collaborative synergistic partnerships.** Using subcontractors, develop partners who share risk, who have skin in the game, and who get paid only through measurable function. Remember to never pay a person as a subcontractor; pay an LLC or a DBA (Doing Business As) so it's a legal business entity, not a person. Otherwise the arrangement will be suspect as not truly a subcontractor.

4. **Create community.** I'm a big believer that unless a farm has two salaries from two different generations it's not sustainable. By that definition, many farmers who claim that mantra actually aren't. Include customers in this as well, but you need people who can cover for each other, leverage different gifts, and share equipment as well as expertise. Some on-farm entertainment dinners may be in order to jumpstart this process.

The fifth fear is MARKETING. How do I get people interested? I can't sell anything. What if people reject me? Isn't it greedy and selfish to hawk stuff? We'll drill down a lot more on this one in the marketing chapters but here are some brief solutions:

1. **Diversify the portfolio.** Value adding and adding other items turn $100 customers into $1,000 customers.

2. **Commission-based partners.** Some people love to sell; turn them loose.
3. **Electronic aggregation.** Move away from bricks and mortar and gravitate toward the electronic retail interface.
4. **Sell common, not exotic products.** What are people buying and using? It's much easier to get folks to buy beef than bison, chicken than ostrich.

The sixth fear is BUSINESS. Isn't business evil? What if I make a profit? What about administration, accounting, insurance? Help! Here are some solutions:

1. **Bookkeepers and accountants are everywhere.** Hire one; they love this stuff.
2. **Categorize everything.** Know your enterprise margins.
3. **Track time and motion studies.** Create benchmarks of efficiency and then put a value on different procedures.
4. **Stay in your unfair advantage.** What do you do well, or what can you offer efficiently, that others can't, won't, or aren't? Lots of times a business is better known for what it says no to than for what it says yes to. Don't be afraid to create a niche of competitive success and then stay with it. Don't get pulled off your game plan by every possible scheme.
5. **Shop around for insurance.** And if you're paranoid about somebody suing you, just keep driving on your expressway commute to the Dilbert Cubicle. You aren't cut out for entrepreneurship. Form an LLC or DBA to separate your business from personal assets. Remember that your chance of being sued is in direct proportion to how much insurance you have. The more you have, the higher your chances of being sued.
6. **Profit is the lifeblood of sustainability; don't apologize for it.** It keeps you in business and attracts young people so as you age you'll be surrounded by youthful vitality. What really upsets people is obscene profits. However, if you don't make a profit, you won't be in business very long. For every bad or corrupt

business there are several really good ones. Remember, we're programmed for negative.

The seventh and final fear is OPTIMISM. How can I be excited about my farm business with corruption in Washington, climate change, and the Kardashians on the cover of *PEOPLE* magazine? Isn't life in the pits? Don't I have to be pessimistic? Isn't that the way you show gravitas? How can I justify thinking big and enthusiastically when collapse and ruin are on the horizon? Here are some solutions:

1. **Have a vision.** It drives everything. Concentrate on what you'd like your community to be like and then make it happen. Answer this question: If time or money were not an issue, what would I do tomorrow? If you deny yourself the opportunity to do that, you're squandering everything that is you. Now go do it.

2. **People love big missions and sacred causes.** You're not just a farmer; you're an ecology masseuse. Good farming is noble enough to attract the best and brightest and that's why you're reading this book.

3. **Broaden your borders.** Read eclectically (we talked about that earlier) and expose yourself to people smarter than you. Remember, Tai Lopez, founder of *The Knowledge Society* and the first official Polyface apprentice many years ago, says that you should try to spend 30 percent of your time with people smarter than you.

4. **Faith eventually trumps fear.** Faith in the Creator's design template. Faith in the integrity and righteousness of food and farming done well. Faith in the ultimate abundance of the earth as an object lesson of divine provision through mercy and grace. What do you have to fear?

I confess that I've struggled with how to get this punchy, hard-hitting list into this book since it contains some redundancy, but I decided to offer it in this abbreviated outline form here. I hope this will become a favorite little section that you'll come back to for broadly jogging your thinking.

Fear paralyzes us, and it's the opposite of a can-do spirit. I hope by now you realize you're not the first to face the obstacles you face or suffer the fears that plague you at night. I can assure you that you're also not in the most desperate situation. Others have been more lost, more destitute, more fearful.

My dad used to say "If at first you don't suck-a-seed, then suck and suck and suck until you do suck-a-seed." He had this whimsical way of making profound truth profoundly funny. In the end, nobody else is responsible for how long you stick with it or the savvy of your dedication and persistence.

Write your vision, your mission, on the bathroom mirror where you see it every day. Put your action plan on the refrigerator door. Cultivate friendships among people smarter than you, ideally doing the things you want to do.

And in the words of another great wordsmith and leader, Winston Churchill, "Never, never, never, never give up." You can do this. I know you can because I've seen too many people in too many dire circumstances do it. So git 'er done.

Chapter 5

Assemble A Team

Whenever someone asks me, "Looking back, if there were one thing you could have done better, what would it be?" my answer is simple, "I would have built a team sooner."

We farmers are the most independent-minded cusses in the world. Perhaps it comes from spending so much time alone in the field. Perhaps it comes from wanting to sculpt our own artwork on the landscape. Perhaps it's because no two farms are exactly alike so dealing with nuances of soil type, slope aspect, wind patterns, and hydrology make us feel isolated in our own unique set of circumstances.

Whatever it is, organizing farmers is worse than herding cats. Teresa and I are cat lovers, more than dog lovers. My favorite ditty that captures the essence of cats is the one I saw on a refrigerator magnet in a novelty story one time: "If cats could talk, they wouldn't." You could almost say the same thing about farmers.

We farmers develop a stoicism about life because it's how we protect ourselves from being pummeled by circumstances beyond our control. Anyone who has farmed for awhile has endured nature's vicissitudes: dry, wet, hot, cold, bugs, diseases, death, predation.

Bucolic pastoralism exists only about half a dozen days a year. But those are the days captured on film and featured in poems and nostalgic prose. The reality is that farming, as rewarding as it can be, has a grim, tough side that creases the edges of all of us who dare to partake.

The farmer epitomizes the iconic idea of the self-made man. Gazing unwaveringly into his horizon, the farmer prepares for the worst, prays for the best, and usually lumbers along somewhere in between. Into this feast and famine, flood and drought, torrent and blizzard, steps a rugged individual hardened by the elements and necessarily defined by stubbornness. People skills are NOT taught in Farming 101. It's about self-preservation. It's about winning in a war against nature's caprice. I'm well aware that many farmers, especially today, are women. I see these same characteristics in them; I don't think this is gender specific.

Farmers don't build teams. Even in ancient tribal and communal agrarian settings, you see individual shepherds, individual farmers hoeing their rice, individual coconut growers harvesting the crop from their trees. It's a vocation of singularity, dominated by isolation and punctuated with occasional festivals and communal work days. Certainly today's farmer is more isolated than even a few years ago when threshing rings, sheep shearing rallies, and butchering parties occurred routinely in the countryside.

These dynamics, most culturally intense within the farming community, are not shared in the greater business world. In any other business except farming, practitioners attend leadership seminars and team building workshops. In the greater business world, success and legacy are all measured by group dynamics and building relationships. Most farmers have a far better relationship with their tractor than their spouse. The two (tractor and farmer) spend more time together. And the tractor never argues.

Farmers aren't into group hugs. Goodness, we hardly shake hands. The closest we come to each other is leaning on opposite sides of a pickup truck to talk about the weather. We don't do corporate rallies with balloons, balls, and other gimmicks to psyche up conference attendees. When we get together, we kick the tires of tractors and peer into augers, seed cleaners, and manure spreaders. The closest we get to people is putting our sons and daughters on the seat of tractors lined up smartly on the trade show floor.

We farmers spend most of our time with ourselves. We see our problems. We see our equipment, our buildings, our hillsides, our trees. We're tied to these things like a mother hen to her chicks. We babysit; we protect; we train; we nurture and feed. We have our responsibilities, and by golly we're proud of the fact that we don't depend on anybody else to take care of our responsibilities. To need others would be to show weakness, to not hold up our end of this dystopian bargain we've contrived with nature--you won't beat me because I'll make sure my responsibilities win.

We farmers suffer from this self-made image, and I mean suffer. It makes us fritter away valuable time on things we're not good at and really don't like. It holds us back from being able to leverage our gifts and talents on the things that make us thrive financially and emotionally.

Anyone familiar with the latest business leadership materials is aware of the *StrengthsFinder* series. The underpinnings of this development strategy is that we make progress much faster by leveraging our strengths than trying to overcome our weaknesses. We've always been admonished to work on our weaknesses as the ultimate goal. "You can do anything you set your mind to," motivational speakers have extolled.

I've got news for you; they're wrong. If I wanted to play wide receiver for the Dallas Cowboys, I could practice all day every day, for the rest of my life, and the Cowboys would never pick me to play that position. They wouldn't pick me to play any position. They wouldn't even send a scout to look at me. No matter how much I set my mind to it, I'd never get picked.

Want another one? I'll never win the New York marathon. I could want it, train for it, get coached for it, but no matter how badly I wanted it, I will never win the New York marathon. I'm a husky build--that's euphemistic for borderline fat. That means I'm strong, but not fast. I almost have flat feet--not sure how flat they have to be in order to be officially flat feet. When others walk across the floor with wet socks they leave these nice little paisley looking prints. Mine look like snowshoes. I don't know if that's medically flat or not, but I do know my feet are a lot more like a duck than a chicken. Folks, no matter how much I set my mind to win the New York marathon, I never will.

The truth is, you can't do anything you want to do. You can't even do half the things you'd like to do. The truth is that all the gifts and talents necessary to run a successful business do not grow on the same pair of legs. While farming is more than a business, it is a business. And businesses require teams in order to really prosper.

The single biggest fear in people, according to folks who study such things, is the fear of public speaking. The fear of addressing people in public eclipses every other fear: heights, entrapment, drowning, robbery, rape, the IRS. Unbelievable. Now to me, I can't imagine that fear. I'm a performer. When I talk in public, my energy level comes up; I get an adrenaline rush; I almost feel an alter-ego well up inside me. I've heard super athletes, especially marathon runners, talk about the euphoria that takes hold right at the point of exhaustion.

That's what happens to me when I'm up in front of a crowd and they laugh at a joke, nod approvingly at a point, clap at a particularly pithy quip--"Man, I've got 'em in my hand. Bring it on, baby." My face flushes and down deep in my soul I feel my transmission kick into overdrive as I head toward the ultimate finish line: a standing ovation. As I give my final line and bow, I'm thinking: "Come on, get up, people. Get up. Come on." And when they do, wow, that's better than gold. When they don't, I have a let down feeling knowing I didn't win this one. Oh well, next time. We'll win next time.

What I've just described will make more people than not break out in a cold sweat and head to the restroom in nauseous spasms of dread. Many people tell me that public speaking wrings out all their energy and they're like a dishrag when finished. Goodness, I've flown into speeches with no sleep, from somewhere on the other side of the world. But when I get up on that stage, I'm as on fire as if I'd just rested for a day. The point is this: each of us has strengths and weaknesses.

The whole idea of partnerships is finding complementary relationships so that one plus one equals three. Do you know how awkward it is to see someone trying to do something they don't like or aren't good at? You can sense the fear and distaste from a mile away. When gifted people unleash their talent on something they enjoy more than anything, it's a beautiful thing; the opposite is dreadful.

The old adage that in marriage opposites attract must surely indicate instinctual self-preservation in the subconsciousness. As we come into the farm, the most careful foundational team to develop is the signature partnership. Normally, that's a husband and wife although it can be any two-person alliance. Certainly singular farmers have survived, but they are rare. In fact, their rarity is why people tell stories about them.

I had an interesting conversation with one of our former interns who had been on his own and was struggling. It was an epiphany for me to suddenly realize I had never been alone, and in the words of Robert Frost, "that has made all the difference."

I grew up here on the farm, middle child of a three-kid traditional Dad-Mom family. When my brother and sister left for happier hunting grounds, I stayed and it became the three of us. Then that wonderful, beautiful lady Teresa entered my life and there were four of us. In fact, one of the most singular things Dad and Mom did to bring Teresa into the Salatin family fold was cut her in as a full and equal partner when we set up Polyface. As the outsider, the new daughter-in-law, it would have been easy to protect the family fortune with a separate arrangement.

The what ifs were many. What if the marriage doesn't last? What if she's a shyster? Do you ever really know your spouse until a decade after you're married? What demons lie behind this pretty lady who seems so capable today but tomorrow might turn into a devil? These are all legitimate questions but Dad and Mom, in their gracious hearts and as a demonstration of their faith, extended full, unconditional partnership to my bride. She had just as much ownership as my Dad.

This illustrates one of the first rules of partnerships: they require shared vulnerability. You can't have a functional partnership; you can't build a functional team, unless you can hurt each other. You cannot have a partnership that can only help and not hurt. Helping and hurting are two sides of the same coin, and that coin is spelled v-u-l-n-e-r-a-b-l-e. That's just one of those relational equations that defines life.

Then along came Daniel, our firstborn, much sooner than we had expected. He was born before our first anniversary. But what a blessing that he came so early because Dad, my rock, my mentor, genius and wise counselor, was diagnosed with cancer the next year and began a six-year battle that ended in February, 1988. Daniel was six, and already big enough to hold a board, fetch pliers, and tote chickens to the scalder.

So just as Dad went down, Daniel came up. I see this as completely providential and can only say, "Thank you, Lord." But the point is that when people look at my success, I'm quick to point out it's not been just me. I've had partners. Parents, wife, children, and now far beyond family. But again, I grew up with this magnanimous idea toward partnerships, watching how Mom and Dad extended their hand to my new bride. I hope everyone reading this--especially the daughters-in-law--can appreciate how profoundly this shaped the team dynamics as Teresa and I embarked on this farming venture.

Within any two-person framework, we have different talents, of course. Dad could fix anything, read blueprints, and loved accounting. I couldn't balance a checkbook, even though I excelled at calculus and trigonometry in high school. Go figure. Dad got bored quickly and wanted to do something completely different and new. I, on the other hand, enjoyed steady progress. I thrived on mowing hay because I could see how much better it was this year than last. Dad was such an innovator that if he mowed that field last year, he wanted to do something different with it this year.

The result was that I brought stability to our team. He had grandiose ideas and I was content to plod along, steadily tweaking and progressing, doing a lot of repetitive actions, but always refining. He liked quantum leaps. I'm still scared of quantum leaps, preferring a more cautious, conservative strategy. Others sometimes tire of my timidity, but I'm quite cognizant of how easily a ship of state can turn bottom up. I don't have Swiss bank accounts to bail me out--or the taxpayers.

Every relationship has compatibility issues. If the two-some is not meshing well, adding more people to the team will not fix the rub. It'll just accentuate it. Here are some common rubs on the team.

1. Starter vs. finisher. To the finisher, that pile of unfinished projects out behind the shed is like a chipped tooth. It irritates and confronts us every waking minute. To the starter, that pile of unfinished projects represents diamonds in the rough, dreams waiting for their wake-up call.

The starter can't figure out why the unfinished piles of projects irritate. They represent progress and the exciting future. "Can't you finish anything?" the finisher asks in frustration. The starter is hurt by the accusatory tone. "You don't appreciate all these good ideas. I'll get to them . . . just you wait and see. And when that gizmo over there is done, life will be good and you'll be sorry you chided me for starting it."

And so it goes. These are the real tensions and practical day-to-day interactions on the farm. Somehow you have to figure out how to come to an understanding that works for both. Both parties must give a little. Just like budgeting is important to stay on financial track, itemizing projects and agreeing where time and effort will be invested is key to harmony. When we live where we work, we can't get away from the business, either emotionally or physically. So all the tensions rise up to remind us every day that they exist. These tensions are in your face all the time.

2. Messy vs. cleany. Some people thrive in chaos; others are obsessive compulsive order freaks. Often these marry and that creates a problem. While this is not a marriage manual, I can assure you that farming accentuates every nuance of marital dysfunction. When both parties head off to their jobs each day and come back together each evening, it creates space for these tensions to ease. But when you work where you play, where you sleep, where you relax, every possible rub point becomes bigger and sharper.

What will the homestead look like? Where will the piles of boards, steel, metal roofing and hand tools be stored? Organization certainly creates efficiency. A place for everything and everything in its place, as the old adage goes, is a good principle for the farm. At the same time, heavy handed addiction to meticulous order can constrict innovation and actually getting things done. We're after balance, not fetishes.

I'm a stickler for some things and completely blind to others. Everyone around Polyface knows that tractors always get parked under roof before nightfall. If it's in four-wheel drive, whether tractor or truck, it needs to be taken out of four-wheel drive when parked. The front end loader is always lowered and both hydraulic directions wiggled to zero pressure in order to keep O rings from wearing out. My ATV has five bungee cords: three on the front and two on the back. No more and no less.

I get pretty anal about some of this stuff. But mowing around the garden? Not a big deal. Teresa, on the other hand, couldn't care less about the bungees on the ATV, but gets put out if waist-high weeds take over the grassy areas around the garden. Most people like things to look nice; our customers want things to look nice, I assure you. But that doesn't require white board fences and manicured lawns. Profitable farms have a bit of a threadbare look because money is going toward more important things. I always tell visitors to look at our fields, animals, and earthworms; that's what's on display. The buildings and grounds aren't where we shine.

3. Spender vs. saver. I don't want to hash over what we talked about in the frugality chapter, but I think it's important to realize that this is another big source of tension on a farm.

It dovetails with the others because the spender who is also the cleany views spending on appearance as necessary. The saver who is also a messy sees it as completely unnecessary and a waste of money. Again, compromises are in order. Generally no two people will have exactly the same view toward money. Identifying the important things and prioritizing projects that cost money will help synchronize these two disparate tendencies.

An ancillary element to this one is the constant tension between the home and farm. Does the house project get the money or the farm project? Do we buy a tractor or put in a grey-water system in the house? Because the house is typically not seen as an income generator but a money pit (hence, the radio show), farmers tend to put spare money into farm projects while the house goes lagging. Having done this most of my life, I can assure you that some of the best money you can spend is on the house.

When Mama ain't happy, ain't nobody happy, right? We can talk about cows and tomatoes and pastures all day but if we come into a contentious house, none of it matters. All parties have to feel honored about getting a piece of the pie occasionally. If they don't, resentment builds. I know because I've been there. And will probably be there again, heaven help me.

4. Introvert vs. extrovert. This is a big one for Teresa and me. One-on-one she'll talk to you all day. But put her in front of a crowd and she's not a happy camper.

So the other day we were eating breakfast down in the kitchen, which is halfway submerged from ground level. I'm at my regular chair at the end of the table and she's at her chair along the side of the table with her back to the main window. I notice some shadows fleeting across the back porch and look up at the window.

A family of four--mom, dad, 10-year-old daughter and 8-year-old son--stoop down against the window, place their hands around their faces so they can peer in, and suddenly one of them screams: "Here's where they live, here's where they live!" Okay, all you folks that want to be like Polyface, try that on for a change.

I look over at them, smile big, wave a welcoming hello. Teresa, with her back to the window, grits her teeth and through pursed lips seethes, "Give me the shotgun." That's a perfect example of how an extrovert and introvert respond to the same situation in extraordinarily different ways. People always ask me why Teresa doesn't travel with me to do speaking engagements. Are you kidding? She's flippin' excited to be rid of me! What are you talking about?

She can finally sleep in, doesn't have to make breakfast, and can stay up as late as she wants. Goodness, if I didn't go away she'd never get anything done. When I'm home life is a whirlwind with media, interviews, customers, visitors. When I'm gone it's quieter and settled. She's been known to hide in a closet. Life is crazy sometimes. I had a guy stop me the other day as I was walking in from the barn. "You work here?" he asked, good-naturedly.

"Yes," I said, nonchalantly.

"Tell Joel you've got a beautiful place here. I'm impressed," he said jovially and headed on out to peruse the brooder house and sawmill. I didn't say anything but laughed all the way into the house. I love greeting people, finding out where they're from, listening to their stories. Teresa purchased a couple of shirts emblazoned with a landscaping service logo. That way she can work out in the flowerbeds and visitors just think she's Maria from Pedro's Landscaping. Isn't that a hoot?

After a particularly hectic season about ten years ago, we decided something had to be done. I couldn't get anything done because I just had to say hi to every visitor who came by and Teresa felt crushed by the spontaneous flow of people. We decided to offer a scheduled Lunatic Tour, limited to 100 adults and the children who go with them, about twice a month. That would give people a dedicated time when they knew they could catch me and it would funnel most of the visitors to specific days. We offered the tour for free.

It worked, but only partly. As the season progressed, fewer people honored their obligatory seats so that by the end of the season, we were down to about 40 or 50 attendees. We'd turned away twice that many saying we were full. We decided we had to ask for visitors who wanted the tour to put some skin in the game. We charged $10 a seat and kids 12 and under were free. It worked. We've increased the price a little over the years but essentially it's the same as a movie ticket or museum fee. I don't know if that means I'm an actor or a relic.

We still get visiting people at other times, of course, but at least a lot is compressed to these tours and it's helped bridge that tension between introvert and extrovert. When you're direct marketing, this tension is more acute. Some people can't imagine that here at Polyface we still maintain a 24/7/365 open door policy. But in reality, nobody comes at 2 a.m. So it sounds worse than it is.

What about phone calls? Many years ago, after a particularly frustrating day, I suggested that we keep a tick sheet for how many phone calls we got in a day. Daniel and Sheri got 50 and Teresa and I got 50. I put my foot down: "We have to do something. I hate coming into my own house because I have half a dozen phone calls to answer and get two more before I can even answer those." This was before email was so popular and before Facebook even existed.

At that point, we did some figuring and decided to hire someone to answer the phone. It was a huge investment, but we needed our sanity back. We hired Wendy, a new mom, who simply answered the phone from her house. It enabled her to work from home and freed us from all the phone calls. She was our first significant non-family hire and one of the best decisions we ever made.

We converted our private lines to unlisted numbers. This compromise enabled us to continue talking with people and letting the informational/marketing end of the business grow, but protected the introverts from having to deal with it.

By this time, the accounting was getting to be too big a task for Teresa. Dad had trained her early on so when he passed, it was a seamless transition because she was already doing the bookkeeping. We hired a bookkeeper to help Teresa. Again, one of the best decisions ever. People are out there with skill sets just waiting to help you. Don't deny them the chance to express their talents.

Assemble A Team

As my son Daniel says, when we refuse to let folks help us, we deny them the chance to express their gifts. Think about the worst thing you have to do today. What job makes you want to stay in bed, to not even get up and face the day? It's that dreaded task. Did you know there is someone, probably much closer than you think, who can't wait to jump out of bed and do that very thing? Amazing but true. Whatever you don't like to do, find the folks who like to do that and add them to your team.

I can hear people choking on that. "I can't afford to pay myself, let alone pay someone else. How am I supposed to have an employee when I'm struggling with my own salary?" This brings us to the commission-based model. If you want the whole scoop on this, read my book *FIELDS OF FARMERS*. The idea is that rather than paying wages, you create performance-based remuneration.

Let me tell you a real story to understand what I'm talking about. Daniel married Sheri in 2002. Here we go again, with that pesky daughter-in-law problem. And let me tell you, she's a feisty filly. She immediately begins poking around looking for a place of service. At the time, we were about two years into the Metropolitan Buying Club marketing idea, up to about 30 families, and Sheri saw this as a growth potential marketing opportunity. "I'd like to take over the Buying Clubs," Sheri told us.

That sounds innocuous, but we didn't know if she could market, if she could organize it, if she could communicate with folks. I mean, she'd only been living here a few months; untried. Oh, we liked her and couldn't have picked a better daughter-in-law if it had been up to us. But business? Hmmmmm.

We didn't want to guarantee her a salary because if she flopped, it would create a host of family tensions. We didn't want to pay her hourly wages because perhaps she would be inefficient. As a policy, I don't want anyone paid hourly wages. One of the reasons we have so many labor disputes in our country is because with hourly wages, workers always feel like they're working harder than they're getting paid, and employers always feel like the employees spend too much time at the water cooler. This is an inherent tension in the workplace.

So with Sheri, I proposed a commission-based compensation plan. To her credit, Sheri went for it. Now, a word to the young people. Sheri could easily have said, "Look, if I'm going to devote my time and attention to this, I want a guarantee that I'm not wasting my time. I deserve some sort of promise; otherwise, why devote labor to this?" Yes, she could have said that. And it would have scared me off. Fortunately, she was self-confident enough and professional enough to jump on the commission idea.

"Sure, that will be fine," she said. Now, several years later, we all know that deep down inside she was saying: "And I'm going to knock your socks off. I'm taking this beyond your wildest expectations."

In a dozen years, Sheri moved that fledgling 30-family, $30,000 blip on our income/marketing screen to 5,000 families and nearly $1 million in annual sales. Her commission check is a serious income. She was willing to risk virtually a no-pay scenario to get a foot in the door and it freed Polyface from the risk of paying for something that didn't pan out.

This is a basic philosophical template for business partnerships, and I'm using the term partnership not in a formal legal sense but in an informal relational sense. Shared risk and shared compensation within the confines of written spheres of responsibility create a balance between autonomy and oversight.

Every team situation struggles with balance on two levels. We see this most acutely with our intern/apprenticeship program. Again for all the skinny on that, read my book *FIELDS OF FARMERS*. I'm not going to go into that here because it's all there and I make it a point not to repeat stuff in my books. That way you have to buy them all because there's no redundancy. See, once a marketer, always a marketer.

Here are the primary tensions we see in team development:

1. Micromanagement (lack of trust) **vs. independence** (too much trust). Perfectly equal partnerships seldom exist; usually some sort of hierarchy defines relative power and authority of the parties involved. Ultimately, functionality requires someone to be in charge, to make the final decision.

When people ask me how Daniel and I get along so well, I quip, "I just give him everything he wants." That tongue-in-cheek response belies a serious undertone, though. Somebody has to wear the big boy pants. Sometimes that's me; sometimes it's him. I've enjoyed watching him take those reins and ride with finesse.

Everyone wants to feel needed, to feel trusted, and to feel like they're contributing to the greater good. Nobody wants to feel like a liability, like a drag on the system. The leadership wants to get people into a higher state of trustworthiness as quickly as possible. Good leaders are delegators. Show me the person who won't release responsibility and I'll show you a business that will never get out of stuck.

When these interns arrive, they don't know anything, so we as leaders must literally micromanage. Before long, they start feeling confident. Of course, we as leaders, as the bosses, if you will, want to back off as soon as possible. We want to give independence and autonomy. But if we give it too fast, we have a disaster. So to protect our business, we provide oversight, which can be interpreted as micromanagement and lack of trust, to our level of comfort.

Our release time and the interns' perception of competence are often not on the same clock. Hence, the tension. We've erred both ways. The tendency is to over-protect because once you, as boss, have picked up pieces from a few disasters, you become more and more reticent to release control. This is just a fact of life. We spend a fair amount of time discussing this balance with the team, using real examples from a real day, in order to explain the tension. As long as everyone understands the tension; as long as the struggle is out there in the open, it's not as volatile. Suppressing the tension's existence creates a powder keg.

2. Working hard (slaves) **vs. free time** (bored). The second team tension involves levels of effort. On a farm, the work is never done. Something else can always be done.

For us, what's been helpful is to have designated supper

times. Breakfast is flexible (we always go out and do chores before breakfast) and lunch is flexible. Rigidity on those meal times can impact efficiency big time because you might have to quit 15 minutes before finishing something. But supper, at the end of the day, is fairly rigid. We try to eat at 6 p.m. when interns are NOT here (October through April) and 6:15 when interns ARE here (May through September).

This creates a definite end to the day. We farmers are notoriously workaholics. I'd rather work than do just about anything. My recreation is moving cows, mowing hay, running the chainsaw. Work is not an enemy; it's a blessed friend. It defines my Joelness, if you will. But if I get caught up in a project and don't come in for supper until 7 or 8, that throws everything off in the house. Washing dishes, organizing the kitchen for the next day. And not to mention, supper is cold and everyone is frustrated.

Mixed farming is a full-on situation, especially during the busy season. In the off-season we spend a lot of time reading, writing (like now), resting, planning and just catching up with ourselves and each other. But during the season, and especially when making hay, it's crazy busy. You can take over-busy for awhile if you know it's not forever. Moving the farm labor effort with the season, letting it ebb and flow, creates pressure and releases pressure. That's a balance we have to confront and maintain if we expect all parties on the team to remain loyal and gung-ho.

With those tensions in mind, let's talk just a bit more about this commission-based team partner arrangement. Rather than having employees, we have Memorandums of Understanding (MOUs). These are simply written agreements that articulate a duration, responsibilities, and compensation. The result is that our team is comprised not of employees, but subcontractors who develop their own fiefdoms. Doesn't everyone want a fiefdom?

Without reiterating *FIELDS OF FARMERS*, let me give a couple of examples of how these work. Jonathon offered to take over the shop and head up Polyface maintenance. We'd never had a shop guy. But he created an LLC business called FarmFix and took over the shop, billing us for repair time just like any professional mechanic in town.

Brie wanted to start school tours. She developed Grasstains tours as her own business. And although she went on to other things after a couple of years, it's still functioning today with another subcontractor. She develops the tour, does the marketing, scheduling, handles payment, everything. Polyface gets a royalty, but she gets everything else. That way if she flops and nobody comes, it doesn't cost Polyface anything. But if she's successful and it takes off, she makes really good money and we get great public relations and potential customers. It's a win-win.

Dan wanted to start a mushroom business. Great. I cut the logs, he inoculated them, cared for them, and sold mushrooms to our restaurants and on-farm customers. When he moved on nobody wanted to take it over so that enterprise died. I don't feel compelled to maintain enterprises if nobody wants to run them. Now I'm freed from being a jobs provider, and instead see Polyface as offering a germination tray to young people who want to create their own enterprises.

These shared enterprises leverage our buildings, expertise, shop tools, equipment, and land. Eric wanted to tap maple trees. Go for it. We lease several properties in the community and these are operated by other young people who build their own compensation packages, their own fiefdoms. They've come through our intern program, which vets them and minimizes train wrecks.

Almost anything can be subcontracted if you're clever enough. I just read today that when Warren Buffet rented his farm to his son, Howard, he offered it at 26 percent of gross receipts for the rent but 22 percent if he kept his weight down to 185 pounds. Now that's creative, tying rent to the weight of the farmer. "Boy, if you get fat, it's going to cost you." Isn't that a hoot?

I'm convinced that the permutations on this theme can apply to far more situations than we think. All of our Polyface marketers are on commission. Our delivery runs on commission--so much per pound, articulated as a separate line item. We pay a certain amount per day per group of cattle for caretaking services. A certain amount per dozen saleable eggs. The beauty of these procedures is that they remove the classic rub between management and labor.

If I'm paying you by the hour to move the herd of cows at ABC farm, I'm going to get frustrated (perhaps even vexed) if you fail to take the proper tools and have to turn around and come back to the shop, in my view wasting a half hour of time due to negligence. But if I pay a set rate, X, for that action, your wasting an hour doesn't frustrate me because it doesn't cost me anything.

If you want to lie down in the grass and commune with butterflies for half an hour, that's fine. It's your time, not mine. But if you're on my clock, I don't want to see you out there communing with butterflies. And I don't want to see you having to come back for the tool tote that you know you're supposed to take. All these tensions ease with non-wage arrangements.

One final point on this. We word all of our MOUs to favor risk taking. The more risk the subcontractor takes, the better the remuneration. We load the compensation packages to the risk side, to encourage risk taking and ownership. Nothing creates responsible actions and attitudes like ownership. The problem with most employer-employee relationships is that if the employee costs the company money, his paycheck doesn't change. That's not shared risk.

In the Warren Buffet case I mentioned earlier, I find it affirming that he based the shared on gross income, not profits. I've never been a fan of profit sharing because figures lie and liars figure. What's profit? You can figure it a hundred different ways depending on what you put in. It's a subjective formula. Do you accelerate depreciation? What's the value of the inventory? A lot of guesswork goes into accounting, regardless of what accountants may say. A lot of assumptions are at play.

But gross income is a set, objective figure. Look at gross income; it's one amount and it's hard to fudge. If your gross margins are good, then more income creates more in-pocket money. Anytime I share income, I want it based on gross not net. It reduces arguments.

Some of the most obvious and rewarding commission-based arrangements involve marketing and sales. Especially sales. With accounting software today tracking sales is easier than ever. Sales representatives can be put on straight commission. This means you have to push a pencil on your margins to tease out what sales cost.

And realize that different products, and even different clients, have different sales values. For example, selling $100 worth of vegetables takes far more effort than selling $100 worth of beef. If the beef is frozen, it's also far less perishable, which creates a stable window.

An institutional buyer like a significant restaurant is more sales efficient than a single family. A $500 sale takes less effort per dollar than a $50 sale. From an administration standpoint, 10 Paypal transactions are far more costly to register than one big transaction. Ditto for checks or cash. The point is that volume sales should be rewarded with discounts both to the client and as a part of any commission-based compensation package.

What is a new client worth? Often marketing headhunters will work for a given amount per client. I know one electronic aggregator that paid $25 per customer and had a whole army of pavement-beaters out there knocking on doors. It worked. They built a business without guaranteeing a single person a paycheck. You only got paid if you brought money in. Believe it or not, lots of people are willing to work this way. In fact, I'd say that folks who don't want to work this way are not the kind of people you want working with you. Anyone not confident enough to work from a performance-based compensation plan is probably not savvy enough to put on your team.

Keeping your team excited and loyal is now your responsibility. Praise often and criticize seldom. My favorite is Stephen Covey's *7 HABITS OF HIGHLY EFFECTIVE PEOPLE*, which is great on this point. He talks about emotional equity being like a gas tank. To maintain functional partnerships, I have to load emotional equity into the tank because sooner or later I'm going to make a withdrawal-- like a harsh comment or inappropriate criticism. He says you have to have enough equity in the tank to make that occasional withdrawal without draining the tank. The problem is that most of us don't put enough equity into the tank and our relationships run out of gas.

You need to provide clarity for the team. Be aggressive about communicating vision, where the business is, successes, and failures. And for crying out loud, ASK your team for advice, for their thoughts. They're out there doing their thing; you can't see everything or be aware of every communication. Empower your people then rely on your people. Loyal, empowered people don't resent your business success; they want to be a part of its success.

We farmers often carry this beat-down inferiority complex, like nobody likes us and everyone thinks we're hillbilly rednecks with hayseeds in our hair. Hold your head up. If you want people to respect you, then respect yourself. That will draw your team to service, which is the ultimate function of a team. Once everyone buys into service toward each other, you have an unstoppable group.

That service starts with the one at the top. More is caught than taught. People will gravitate to an attitude and action that they see demonstrated by the leaders. Don't like your team? Look in the mirror. Maybe they're developing an attitude and ethic that's just like yours. Ask them to help you change. Serve if you want to be served.

Teams are wonderful when they function. They're worse then being alone if they don't. But they're the secret of sustainability. I'll close this chapter with this strong statement: until a farm employs at least two salaries from two different generations, it's not sustainable.

Lest anyone think I've just stepped onto the slippery slope demanding farmers to be empire builders, I submit that two people do not an empire make. This is a long way from an empire. I'm just talking about a functional, sustainable business. If your farm only employs one person, in the words of Stan Parsons, founder of Ranching for Profit schools, you just created a job, not a business. I'm not suggesting you have to be big; I'm suggesting that everything has a critical mass. Even compost piles need to be a certain size in order to function.

Assemble A Team

Many of the labor conundrums facing farmers could be solved if they'd just think a little bigger, just big enough to bring on another pair of hands. Otherwise, how do you go to a wedding out of state? What if you get sick for a couple of days?

What if you hurt yourself and can't get outside for a week? These are all realities of life and it's foolish to think you'll be immune from personal needs or maladies.

Let your imagination run as to how you can bring on a partner. What size does your farm have to be in order to create another salary? Do it with non-salaried partners and chances are you'll have no problem finding a motivated, enthusiastic entrepreneur who can either grow what you're doing or add a complementary enterprise. In either case, you'll create something with some labor stability and a business with longevity potential. Otherwise it'll just follow the trajectory of your aging process, and that's not a pretty business trajectory. It's certainly not sustainable or regenerative.

Anyone who fears bringing on another person should look deeply into the bathroom mirror and ask yourself, "Do I want to grow old alone?" If the answer is no, then a team is your answer. If the answer is yes, then you're hopeless.

A great metaphor on this point is a fire. You can't sustain a fire with one coal. You need new wood and a couple of coals. That new wood is the new generation and additional coals are team members. I don't know how large a fire has to be to keep functioning, but I know that one coal does not a fire sustain. A functional, successful business requires a certain scale in order to regenerate itself, to sustain people, services, and products. It doesn't have to be huge, but it has to be big enough. That varies from area to area and farmer to farmer. An individual farmer, alone, cannot sustain a business.

Here's to thousands upon thousands of farm teams, a repopulated rural America with land caressed by loving stewards dedicated to healthy soil, pure water, and clean air. I think we can all get on that wagon. Welcome aboard.

Chapter 6

Direct Marketing – Why

Farmers tend to break out in cold sweats and head for the exits when the discussion turns to marketing. We have a built-in fear and distaste of marketing and I think I know why.

Remember, many of us are farmers because we really don't like people. Marketing sounds an awfully lot like a people-centric pursuit. People are far more relationally problematic than cows and corn. No matter what kind of mood you're in, the corn never talks back, makes snide remarks, or punches you in the nose. Cows are always glad to see me. Even if I don't want to see a person, I enjoy seeing the cows because they're never prejudiced or conniving.

Certainly we farmers can get aggravated at miscreant cows or fickle tractors, but we can sell either and be done with it. People? Hard to sell. That aggravating spouse, customer, or employee is there today, tomorrow, next week, next year. Sheesh! Because farming for the most part is a non-communal existence, it doesn't attract or cultivate people skills. Marketing requires people skills.

And not only that, it requires people skills toward people not like us--those pesky consumers, the non-farmers. Those city people. Those urbanites. Those ignorant people who don't know the horns on a cow don't make her a bull, who don't know that a rooster is not necessary for a hen to lay eggs, that salsa doesn't grow on trees and that breaded meat patties in the shape of Dino the Dinosaur is not a muscle group on any critter.

You mean we have to love *them*? *Those people*? Oh good grief, Charlie Brown! Anything but that. Aaargh! Isn't that what Peanuts characters say when everything is messed up? In children's books about farmers, you see farmers planting seeds, feeding animals, and driving tractors; you don't see them on the phone talking to chefs or developing websites to interface with urban customers. And you don't see them butchering anything, canning anything, or cooking anything. You don't see them in the dark loading a van to make a mad dash into the city to service customers and try to get home in time for chores. That's not in the children's books because it's not what farmers do--or so goes the narrative.

You see, farmers are supposed to be colonial serfs, or feudal peasants, doing their solitary planting and husbandry chores, while processing, distribution, and marketing all get done by clever entrepreneurs in the city. Every farm gathering I've ever attended eventually begins demonizing the notorious middleman. Farmers spit the very word out of their mouths with the venomous distaste of the most despicable enemy they can imagine.

Worse than a bureaucrat. Worse than a criminal. "The middleman makes all the profits" is an axiom in farm country. It's part of the unwritten DNA among farmers. They're born hating these middlemen that siphon off all the value. In the 1950s, farmers received on average nearly 40 cents from every retail dollar. Today, that percentage is below 9 percent and trending downward. That means we could establish a new U.S. food policy that farmers had to give their production away for free and it would only change the price of food by less than 9 percent.

If you look at the trend lines in farmgate value versus retail value, their trajectories continue to diverge. Most of this erosion in the farmer's share of the food dollar is not due to eating different foods. It's primarily due to the erosion in domestic culinary arts.

The growth of convenience, pre-prepared, pre-packaged, means that the average consumer is contracting out services previously done in the home. Those services come with a price. Notorious middlemen are more than happy to accommodate this new desire for convenience with its requisite warehousing and distributional logistics. They're more than happy to cook it, slice it, dice it, season it, stabilize it, color it, tenderize it.

And yet as much as farmers deplore these parasitic middlemen, we'd rather demonize them than join them. Why? I can think of four fundamental reasons why.

1. Self-promoting. American agriculture policy can definitely claim one major success: creating a mystique among farmers that the world depends on us to keep from starving. "We feed the world," proclaims any industrial orthodox speaker, press release, convention banner.

The unspoken but obvious conclusion of this mantra is that if American farmers don't fulfill their end of the bargain by planting "fencerow to fencerow," as admonished by former Secretary of Agriculture Earl Butz, we won't be doing our part of feed the world and help America's balance of trade deficit. This obligation to grow things bigger, fatter, faster, cheaper hangs around the farmer's neck. Failure to comply means we farmers cheat the world from the wealth and benefit of American production prowess. After all, the world is depending on us to feed the starving children in Africa or IndoChina. I don't want to head down a political rat hole with this discussion, but somebody needs to say directly and firmly: "We're all responsible for feeding ourselves."

That is not uncharitable; it is practical and the responsible place to start. When you see what permaculture can do, for example, to rehydrate sub-Saharan Africa, or what controlled grazing can do in South Africa, you begin to realize quickly that every society ultimately has to figure out its own food issues. Every society and every place has solutions. Virtually every time food shortages occur in an area, they are not based on resource limitations, but on social and political weaknesses.

For the American farmer to be held in emotional extortion to these dysfunctional cultures (as if the American is not dysfunctional--ha!) is unfair. We all have our own problems and must work at solving them. Of course philanthropy is good, but often handouts actually inhibit creativity and a can-do spirit.

Nonetheless, many of us farmers feel mercenary and selfish when we start hawking our own wares. To self-deprecating farmers burdened with this "feed-the-world" cloud hanging over our heads, marketing self-promotion seems awkward and prideful. Can you think of any other vocation in which good practitioners would feel awkward marketing their product or service? Does a photographer owe the world photographs? A carpenter furniture? A builder houses? A plumber pipes? An electrician lights? A writer books? A banker money?

Thinking like this for any other profession is absurd, and yet farmers are supposed to think of ourselves this way. It's ridiculous and too often keeps us from getting out there and stirring the marketing pot. If you produce something great that can help people be healthy and well, like good food, you should be encouraged to promote it and explain it to people. So hold your head high and get over the "feed-the-world" shackles. Feeding the world doesn't pay your bills, stop your gullies, keep the bugs away, or water your squash. If you're going to do that, then you deserve to tell your story so others can appreciate what you've done.

With all that said, more often than not the marketer is someone different than the farmer. Not only does marketing take a different skill set, having another person do it creates a bit of separation between the producer and the promoter. Be assured that when I extol the virtues of marketing and say you need to do it, I don't necessarily mean you the farmer; I mean you the organization. Someone in the farm organization needs to do marketing, but often that won't be the primary production person. It's a facet that must be done in order to thrive. But it doesn't have to be the primary farmer. Okay, now you can breathe again.

2. Emotionally vested. I could call this the fear of rejection. As farmers, our products truly are an extension of our personhood. My chickens are not just any chickens; they're MY chickens. The ones I fed, watered, sang to and protected from the weather and the predators. The ones I watched eat grass and bugs. I chose chickens instead of radishes because I like chickens; they nurture my soul. If I were a radish guy, I'd grow radishes, but I'm a chicken guy. I like chickens.

The point is that what we grow is our life affirmation. If you don't like my chickens, you're not just rejecting my chickens, you're rejecting me. When a customer doesn't buy our product, we farmers get all weepy inside and begin to cry, "You don't like my chickens? Do you know what I did to bring these chickens to you? I babysat them when they didn't have a mother. I adjusted the heat on the brooder and checked them in the middle of the night to make sure they weren't too hot or too cold. I wiped their bottoms when they got pasty and dipped their beaks in water when they looked listless. I played Beethoven to them and read poetry."

Okay, you get the picture. Many farmers don't want to even try direct marketing because we fear the emotional fallout of rejection. We're too close to our products. We're not just one little piece of a bigger product; we're the whole enchilada. We live and die with our plants and animals. They represent the sum and substance of our being. And if somebody doesn't salute our carrots, we take it personally, as if the non-buyer beams rejection energy at us.

Let me ask you something. Do you beam negative thoughts at the brands you don't buy? People choose to patronize products or services for any number of reasons. It could be because it's the brand Mom used. It could be we've invested in that particular company. We have lots of reasons for choosing one product over another, and seldom do we channel evil thoughts to the ones we don't buy. We just don't buy them. It's a big world with lots of choice.

Farmers who market must cultivate enough self-confidence and personal affirmative identity that we don't melt when somebody turns down our peas or pork. Amway (now Quiksilver) fights this problem by turning the turn-down into a positive. If I remember the Amway pitch, it's something like this, "The sooner someone tells you they're not interested, the better because then you're freed up to go on to the next potential sign-up. Be glad when someone says no because you can now focus your attention on a new prospect." That's really quite a wonderful way to look at it and an attitude farmers should adopt when selling our produce.

3. Hard work. Anyone who thinks marketing is a cakewalk has not done it. People look at me and think marketing comes easy. Okay, I'll admit that I'm a fairly good storyteller, have a lot of self-confidence, and don't mind telling people I have the best chicken in the world. I've even been known to say that if the Colonel had had our chicken, he'd have been a General.

I can even admit that marketing comes easier to me than most. Fair enough. But believe me, a day of marketing is still much harder than running the chainsaw all day. I think it's because you're matching wits with other people. Running a chainsaw has its own thinking to it, for sure, but it's primarily physics. Trees fall the way they lean. Logs always fall down hill. But in marketing, logs sometimes fall uphill. It's the craziest thing.

Marketing requires communication skill, and that's not on the job description for farmers. The art of negotiation, of making the deal, is not part of the normal farmer's average day. It's certainly not promoted in agriculture curriculums or degrees in animal science. As a result, marketing is the job that gets pushed aside. It's far easier to go out and jump on the tractor than it is to make those two phone calls to potential chef clients. That's even true for me, and I'm not a bad marketer. So if it's true for me, I can only imagine what it's like for most farmers.

4. Peer dependency. Interestingly, for all our independence, we farmers sure do care a lot about what the neighbors say. We sure tune in to what they say about our farm down at the feed mill or auction barn. Farmers are such a subset of the culture now that we're literally a fraternity. One constant that permeates fraternities is the "grape vine" method of communication.

He said, she said, he did, she did. Things whirl through fraternities like a glorified rumor mill. Many farmers can't innovate because we're concerned about what the neighbors will say. We go to the same church or belong to the same civic organizations as our neighbors, so the multi-layered relational overlaps accentuate this peer dependency.

If I go out and tell potential customers to buy my calves, suddenly that puts me in a judgmental position over the neighbors' calves. What happens at the civic club when I tell Peter my grass finished calves are what he should buy and Farmer Paul with the feedlot overhears the conversation? "Are you criticizing the way I raise my cattle?" he challenges, hackles and voice already raised.

To keep the peace and avoid conflict and contrast, we'd rather just not market. Better to go to the grain elevator, the sale barn, the processor with commodity product and avoid all this turmoil. Who wants to be shunned by their neighbors?

This point really came home to me a few years ago when we convinced a couple of brothers in the community to sell us their calves. We were trying to buy all our calves from nearby farmers directly rather than through the live auction barn. When we got these two brothers to sell us their calves, we thought we'd finally penetrated that fraternity. You see in our local good ol' boy farming community Polyface is a typhoid Mary, a pariah. That's why I wrote the book *THE SHEER ECSTASY OF BEING A LUNATIC FARMER* several years ago. I call it my soul book.

I think we bought about 100 calves from them. A month later they called us and asked us if we'd buy another heifer. They had saved back a group of heifers for replacements and had the vet out to soundness check them. The vet said this one gal would be a problem because her pelvis was too small. The best pro-action was to not breed her. We had finished buying our group of fall calves and didn't really want another one by itself to train to electric fence and get acclimated and asked them why they didn't just take her down to the sale barn.

They explained that the folks at the sale barn had black-balled them because these guys did not bring their calves into the sale like good little boys. This is the fraternity. "We just can't face those sale barn guys," they pleaded. I couldn't believe it. I would never think that way. But that's normal in the farming community. It's such a tight-knit fraternity that anything out of the ordinary is like "Go to jail. Do not pass go. Do not collect $200." These fellows had dared to do something different by fraternizing with the local weirdo--me. The backlash from the orthodox farming community had these guys scared to death. Talk about emotional extortion.

Farmers who dare to market must realize that branding and differentiation are the lifeblood of product loyalty. Absent that and you've got what everyone else has. Yawn. Who wants to buy the same old same old? And how can you charge a premium for the same old same old? By definition, marketing is about setting yourself apart, telling your unique story.

These are the four things--all legitimate--both conscious and subconscious, that play in a farmer's mind when the word *marketing* enters the discussion. I understand them all. I've wrestled with them all. I get it. This is why farmers would rather demonize the notorious middleman than become one. However, all these four fears aside, farmers should embrace the idea of becoming the middleman.

Here's why. Imagine a four-legged stool. The round top we'll call a retail dollar. That's what a customer sees and buys. The four legs we'll label production, processing, marketing, and distribution. Those four things have to happen in order for that stool top to exist. Absent those four legs, you don't have a top to sit on.

For sake of discussion, let's assume that each of those four legs contributes a quarter to that retail dollar. Certainly different commodities have different ratios among those four things, but let's just keep the object lesson simple. The principle does not change even if the various ratios do. The farmer, of course, is represented by the production leg. The other three legs are those notorious middlemen.

What do all farmers the world over get together and whine about? Go ahead, take a guess. This is not hard. Yes, you guessed it.

Weather, price, pestilence, and disease. Farm conversations start and stop on these subjects. These are the things that make farmers think we're special and deserve special consideration like subsidies, crop insurance, and grants. After all, we're different than other businesses because these big variables are outside of our control.

When a plague of grasshoppers comes through, how can we stop it? When the rains stop, how can we start them? When soybean rust or whatever makes its way into our region, we can't put up a fence to exclude it. When Africanized honey bees began invading, what could an apiarist do? When Brazil has a bumper crop of corn and the world price drops $2 a bushel, what can I as a farmer do about that? I can't affect that price anymore than I can affect the weather.

This helplessness creates a mystique in the farming community that we're not like any other business and therefore we're victims of circumstances. Society should bail us out because well, well, well, we just deserve it, that's all. Because we're special. And if you don't believe it, just sit down here and I'll tell you how special farming is and why we need price supports and conservation payments and state-run tree nurseries and cost-share programs and yada yada yada.

The average farm, which doesn't do any marketing (signing a contract with the milk processor or the grain elevator is not marketing), is dependent on only one leg for all its income. That leg is production, and that's the most vulnerable leg. When the grasshoppers come, they don't eat the tires on the delivery truck. When the drought occurs, dry heat does not destroy the wi-fi connection to our customers via the internet. They can still see our website even though everything is turning brown outside.

The point I'm making is that not only is a one-legged stool tipsy; if it's the production leg it's also the most risky of the four. Dear farmer, if for no other reason than to build resiliency and stability into your farm business, consider income streams from those other three legs. Every dollar we derive from those other three legs is less vulnerable to the vagaries that have plagued farmers since the beginning of civilization.

Direct Marketing – Why

Look, if the middleman makes all the profits, I want to be one. Sign me up. Pick me, pick me. Rather than demonize, let's co-opt. Don't fight 'em; join 'em. One of the things my dad, as an economist, taught me early on was that as small farmers, we couldn't compete at the commodity level.

A lot of people who believe like me don't like to admit that extremely large farmers actually make money. For a moment, forget the subsidies and the externalized costs. Just looking at the business, expenses versus income, a large farm can be profitable even at extremely low margins. Why? Because when you spread the overheads over that much volume, it works. Even though farmers like me can't imagine buying a $500,000 combine, if that combine runs nearly 24/7 for 6 months of the year covering an average 150 acres per day at $20, that's $3,000 per day for 180 days, or $540,000. When you look at it that way, it works.

But if that combine only does 5,000 acres, it only generates $100,000. Now it's getting dicey. And if it only does 1,000 acres, it only generates $20,000 and there's no way in God's green earth that you can justify a $500,000 combine if it only generates $20,000 a year. The commodity system rewards volume and scale, and strictly economically speaking, it works.

But my dad, as an economist, always said that as small farmers we can't play the low margin commodity game because we can never grow enough. Oh, we can try to get by with a $10,000 combine and a truck held together with bungee cords and epoxy. But in the end the big boys will grind you up because they are chasing the price to the floor and as a small producer we can't spread our overheads over enough stuff.

The same thing is true in livestock. Stan Parsons always used to say that if you are using more than one salary per 1,000 cows, you won't be profitable. The salary is the same whether that position services 50 cows or 1,000.

But as soon as you start wearing those other hats, the hats of the middleman, everything changes. Now you don't have to chase that commodity cycle; you simply create an entirely new economic paradigm. Not only do you begin insulating yourself from the farmer's typical worries and vagaries, you also turn each production unit into a much higher value. This is the whole idea behind value adding, which is a hot topic in small farming today.

A head of cabbage might be worth $1.50. But in artisanal sauerkraut, it might be worth $10. If you need to generate $100,000 to pay overheads and your living expenses, you can do that on 66,666 heads of cabbage at $1.50 or 10,000 heads of cabbage at $10. It doesn't take a rocket scientist to realize that it's a lot easier to find land to grow 10,000 heads than 66,666. And it'll be easier to keep the bugs off the 10,000, make sure they get water, apply the compost, plant, harvest, and look at them all.

This is why time isn't money. It's why volume isn't money. It's why only money is money. Your goal is not to feed the world or get your picture on the front page as the biggest farmer or the highest producer per acre or the one with the heaviest calf at the feeder cattle sale. Your goal is to live to fight another day. It's always to live to fight another day.

Gardening guru Eliot Coleman, in his spellbinding presentations, says that every day the gazelle gets up and hopes it can run one step faster than the lion. And every day the lion wakes up and hopes it can run one step faster than the gazelle. Folks, forget the spots, wrinkles, tail length and whatever; focus on staying alive. Staying alive means we often have to wrap our lives and our arms, our farm businesses, around unconventional income-producing elements. That's how you stay alive.

This whole idea really came home to me the year we began offering cut-up chickens. We didn't want to do it. Philosophically, I was opposed to it. Our vision is to get more people in their kitchens, working with food, learning about food, touching it, getting acquainted with it. We want folks to buy unprocessed, as raw as possible, and then develop it into table-ready fare. But as the years wore on and our customers became more and more ignorant in the steady slide away from the home kitchen, we heard more frequently the lament: "Well, I'd buy a lot more if I could get boneless, skinless breasts."

We didn't want to cut up chickens. Did I make that clear enough? Like the fellow in Dr. Seuss' *GREEN EGGS AND HAM*, we did not want to do it on the day of processing. We did not want to do it in the night. We did not want to do it "here or there," we did "not want to do it anywhere." We did not want to cut up chicken.

But we did. And in the first year, we made an extra $20,000 on chicken without raising a single additional chicken. That's almost another salary. And our customers loved us. At $14 a pound, do you know how much value one 48 quart cooler can hold if it's full of chicken breasts? Easily 50 pounds; that's $700 in one cooler! We'll talk about the economics of it more when we get to time and motion, but the sheer power of value adding is incredible.

Taking everything to its highest retail value allows you to get off that production treadmill. The undifferentiated commodity pricing structure is like a treadmill because all players are trying to become least cost producers. It works as long as you can continue spreading your overheads over more volume. But if you can't get more land or bump the limits of production, the treadmill becomes enemy number one.

Donning that middleman's hat is incredibly liberating and empowering. Becoming a middleman through value adding enables that escape from the commodity treadmill.

As if that were not a compelling enough reason to become a marketer, here are some others.

1. Community economy. One of the secrets of Amish communities is that every dollar turns over seven times before exiting the community. At the top is the retail cheese dollar. But that dollar flowing in circulates through the community from farmer to harness maker to horse breeder to barn builder to lumber miller to buggy maker and beyond. This is the historically normal way local economies worked.

It's why big box stores never actually create stable communities, regardless of how much they give to the United Way fundraiser. In the end, locally owned and locally operated businesses are the underpinnings of strong local communities. As I'm writing this, Teresa just walked by my desk to show me the wedding gifts she purchased this week. You know how weddings come in triplicate, it seems.

She purchased three beautiful hand-thrown bowls from a neighbor artisan potter whose work is world-class. We actually look for ways to patronize local artisans like this. Who needs Wal-Mart? I haven't been in one in years. The money for that bowl came into the community via our customers, most of whom live outside our community. We like to view ourselves as the reverse economy, bringing urban dollars back to rural America.

2. Attracts the best and brightest. If we need one thing on our farms today, it's our best and brightest. Now don't get me wrong. Plenty about production farming requires skill and knowledge. Good farming especially requires lots of information, lots of know-how and innovation. But marketing is wit-matching, theater, graphics, art, social media, and communication.

Marketing requires a skill set far beyond planting carrots and moving cows, as important and sacred as those things may be. The business acumen and savvy required to develop those other three legs of the stool are difficult and challenging enough not only to attract our best and brightest, but also to keep them interested for a lifetime. We need farms and farmers inhabited and stewarded by the absolute best people our country can put forward.

The old idea that dummies farm has created rural brain drain and placed our most valuable air, soil, and water resources in the hands of D students. We must reverse that trend. I suggest the way to steward our air, soil, and water better is to put more clever and thoughtful people in charge of them. The magic in school gardens and backyard gardens attests to the instinctual attraction in the heart of every human to connect with our ecological umbilical. I'm convinced that thousands upon thousands of young people would embrace a farming vocation if they thought they could actually make a decent living at it.

That such a large number of capable and innovative young people are being told by parents, guidance counselors, professors and friends that a farming vocation is both demeaning and impossible is a blight and curse on our culture. For a civilization as creative as ours, imaginative enough to develop the internet and the shared economy, to be unable to offer life-satisfying attraction to the most fundamental vocation of life--growing food--indicates a profound bankruptcy of value.

3. Customers will move with you. Later we'll drill down into the concept of the portable farm, but for now, let's appreciate that customers are portable. They will move with you if your circumstances change. In today's business parlance, they're nimble.

If the grain elevator goes out of business or if you move operations somewhere else, that stationary buyer won't come with you. If you downsize, upsize, turn right, or turn left, your customer base will hang in there. When people ask me the most valuable thing on our farm, I always point to our customer box that contains their names and contacts. Most of that is on the computer now, but that list represents our true equity.

Customers are better insurance than the insurance company. They're better investors than the bank. They're better helpers than bureaucrats. The scale-dependent commodity business relies on highly capitalized single-use infrastructure customers. When you become a marketer and begin taking your product to the final end user, you spread your market over lots of people.

I'm reminded that even in the 1929 great depression, unemployment was only 30 percent. For sure, that's a high number. But it means employment was 70 percent. That means that even in a depression as bad as that--and many think we'll have one again before too long--70 percent of folks never lost their jobs. So if you have 1,000 customers versus one grain elevator, one processor, one co-op, or one sale barn, chances are 700 will continue to be your customers even in the worst of times. Folks, that's resiliency.

4. Emotional support. A farmer's life is often difficult and lonely. To have an appreciative constituency is literally emotional life to the farmer. Loyal customers want you to be successful even more than you do. I've been given money, time, talent, and automobiles, just out of the appreciative goodness of customers' hearts.

Believe me, if you drop out of your processing contract or don't show up at the sale barn one year, nobody's going to come out inquiring about your health and success. But with direct marketing, you garner to yourself an entire cheerleader squad, rooting for you and encouraging you on the sidelines.

This is important too with your children. Today farm children often receive condescending jokes from schoolmates. But when your customers interact with your children and applaud what they do with comments like "our family depends on you for our lives. You guys are awesome, producing the best stuff in the world. We just love what your family does;" do you know what that does to your kids?

They strut around and think Mom and Dad are more important than the mayor. Such affirming interaction instills in them a love of what the family does, a love for what the farm can mean in the community, and a spiritual bond with this sacred vocation. That's priceless. In my view, ultimate sustainability is about what the next generation does.

Certainly farmers do not want to put our children into bondage by requiring them to farm, but we should create a habitat that engenders a love of farming in their psyche. One way to do that is with marketing. Our direct marketing customers will be the first to patronize the children in their entrepreneurial activities.

I'm convinced that many entrepreneurs don't develop simply because they did not have a germination tray early enough in life. Marketing, with a client base interacting with your farm, offers that germination tray for children to ply their entrepreneurial trade. It may be something related to farming and it may be something totally different. But how many families have 1,000 patrons routinely interacting with their children?

That's a powerful motivator for the next generation. Parenting is hard enough without help. But having all these affirming, encouraging folks coming around helps direct and develop our children in profound ways. Through this interaction, our children learn social skills, self-confidence, and how to negotiate and interact with an eclectic blend of people. In doing so, they also develop a wonderful sense of respect and honor to different kinds of people. They learn early that dark skinned and light skinned and people with accents and people who wear odd clothes--all these people appreciate what the farm offers and what the family does. That is the best place I can imagine for kids to grow up.

Why should farmers market? For all these reasons and more. These are just the ones I'm putting down here. Marketing has a host of benefits. Marketers own customers. That's the best investment you can make.

Your Successful Farm Business

Polyface in Photos

Photography by Jean Shutt

Polyface in Photos

Polyface Central. A hub of operations needs to be organized and inviting for visitors.

Polyface Broiler Shelters. Mobile, modular, management intensive.

Polyface in Photos

Pastured Poultry, Polyface style. Inserting a dolly enables one person to move 5,000 chickens per hour without starting an engine or burning petroleum. The chickens walk on the ground to the next spot.

Pulling from the other end, moving broiler shelter. This enterprise can be stacked on almost any farm in the world.

Poultry processing shed at Polyface, known as the disassembly line.

Processing turkeys for Thanksgiving.

Polyface in Photos

Eggmobiles follow the cows in their pasture rotation. Chickens sanitize the paddocks. The portable hen houses come in all shapes and sizes. Top shows 12 ft. X 20 ft. and bottom shows 8 ft. X 16 ft. Engineering is easier when two are hooked together than trying to make it all one big trailer.

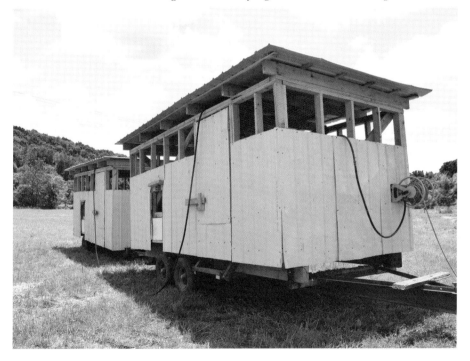

Heather Juda built her own fiefdom hatching chicks for Polyface on land Polyface leases for grazing cattle. The seedstock criteria are simple: old, healthy, and productive. The chicks are bulletproof.

Granddaughter Lauryn Salatin shows off one of her exotic pullets named Rockstar, a Silver Polish chicken. Her business is buying chicks and selling them ready-to-lay to folks who want a rainbow backyard flock.

Grandson Travis Salatin's duck enterprise, producing eggs from portable pasture infrastructure called the Quacker Box.

Grandson Andrew Salatin's sheep. Start them young as entrepreneurs and give them autonomy.

Raken House (Rabbit-Chicken). Perfect example of permaculture type stacking for symbiotic hygiene and economics.

Son Daniel moving a Harepen. Mowing with animals saves petroleum, fertilizes the soil, and generates income. What's not to love?

Laying hens and rabbits come into hoop houses for three months in the winter. Deep bedding, sunlight, and symbiosis provide sanitation. Stacking creates efficient use of space and extra profit.

Allison Hayes, former intern, tends vegetables in the hoop house as her fiefdom once all the animals go on pasture in the spring. Multi-use infrastructure generates more profit and reduces pathogens.

Millennium Feathernet at Polyface, an X-trussed A-frame on skids, moved every three days and big enough to accommodate 1,000 pastured laying hens.

Electrified poultry netting is the technology that allows large commercial flocks of poultry to enjoy more sanitation and chicken-friendly habitat than a dirt yard flock in the typical American homestead of the mid-1900s.

Polyface in Photos

Gobbledygo with turkeys. Note shade cloth and perches. Everything is notched to enable efficient dismantling in order to be highway-legal. A perfect example of the portable farm.

Nature's biomass pruner, soil builder, and pastoral landscape painter: the herbivore known as grass finished beef. Choose phenotypes that fit.

Battery, energizer, and ground rod set-up for electric fencing. This is all you need for internal control.

K-line irrigation from New Zealand was designed for small acreages and misshapen fields. It's a perfect insurance policy for agriculture. Water is from stored winter runoff, not aquifers or streams (the commons).

Polyface in Photos

The Polyface band sawmill is foundational for generating lumber, value adding timber, and as another profit center. Note the sign: Joel's Man Cave.

Daniel and apprentice Chris Slattery feed branches into a Valby Chipper as part of forestry operations at Polyface. This is the carbon economy in action.

Access through a valley can become an asset if you build a pond and put the road over the dam, creating an example of multi-purpose working landscapes.

Pigaerator compost in the carbonaceous diaper. The cow hay feeding system moves vertically to accommodate the bedding build-up. Fermented corn pays the pigs to aerate the pile. This is Polyface fertilizer in the carbon economy.

Pastured pigs, controlled with electric fence, nested in the landscape, moved frequently to avoid overly heavy impaction.

A Lunatic Tour stops off at the eggmobiles. These 2-hour hay ride tours create their own income stream and promote both information and product.

Polyface farm family dinner: interns, apprentices, staff, and owners, eating communally. Substituting people for drugs, energy, and capital-intensive infrastructure moves equity to skill, information, and customers.

Chapter 7

Direct Marketing – Where

Most of us have heard marketing gurus, especially retail venue experts, say, "The three most important things to remember are location, location, location." We've talked about branding and messaging, but it's all for naught if we don't participate in the right venue.

To the farmer, we have a host of venues from which to choose. Let's go through them with pros and cons.

1. Big Supermarkets. "When can I get your stuff at Wal-Mart?" folks ask me. My first response is that if I don't go in their front door; why would I want to go in the back? But the more complete answer is that it's a big-boys, empire-building relationship that simply doesn't fit with small farms.

Several years ago the nine vice-presidents from Bentonville, Arkansas came by Polyface as part of a two-day sustainability journey. They'd been out to the Chesapeake Bay and wanted to stop by here as part of their edu-tainment outing. I took them on a farm tour and we had lunch in the sales building.

It was all jovial and they were wonderful folks. As we were eating, one of them asked me point blank: "So how do we get

Polyface products into Wal-Mart?"

I responded, "First, you need to allow a truck smaller than a tractor trailer to back up to your loading dock." The whole room went silent and they all looked at each other as if somebody had just dropped a live grenade on the table. The conversation quickly turned to other things and they never brought it up again. You see, they couldn't conceive of a business model consistent with theirs in which anything smaller than a tractor trailer would back up to their dock.

Some might admonish me for thinking too small and suggest that farmers like us should band together to get enough volume in one place in order to fill a tractor trailer. Assuming that could be done, look at the effort. Just getting the product liability insurance that would satisfy the supermarket underwriters is an expensive and protracted proposition. The insurance underwriters are going to ask for everything from compost licensing to wildlife exclusion (we can't have duck poop falling on the green beans or a deer running through the radishes).

Everything is getting squirrely today as the industrial orthodoxy heads into sterilization scorched earth policy in the name of food safety. Chlorine residue on the lettuce is not a problem; an errant aphid is. As if eating an aphid is worse than ingesting milligrams of chlorine.

I called the grocery store Kroger once when we had extra eggs. I could never talk to a person. Another grocery store, Whole Foods, kicked us out because our paper pulp cartons--as opposed to plastic--were not glitzy enough for their shelves. Supermarkets are notoriously slow to pay and highly price sensitive. I've listened to horror story after horror story from folks who went to great effort to comply with paperwork and insurance in order to get into supermarket venues only to curse them a couple of years later. These big buyers will jump ship over a penny. It's just not a good fit for anyone who wants to market quality more than quantity.

And now that Wal-Mart is the world's largest organic vendor, it's further confusing and complicating the marketplace. This is industrial organics, and it's now a much more insidious enemy to integrity than Monsanto ever hoped to be. I have no desire to dance with the big supermarkets because they're not good partners. Instead of investing time and energy trying to figure out how to get into them, I've found it much more productive to create a completely alternative venue. If one of them actually called me and wanted to do business, I'd certainly consider it as one part of my marketing portfolio.

Supermarkets have only been around since 1946. Why should anyone think they'll be around in 2046? Right now 60 percent of media entertainment material is sold on an electronic platform. This is why Blockbuster went out of business selling and renting videos. Internet sales are taking over physical cashiers as a retail interface. Food is around 2 percent electronic sales. What happens if and when food hits 60 percent? More about that later, but the notion that we need large brick and mortar retail interfaces to move food is burying your head in the sand.

Many supermarkets are now requiring GAP (Good Agricultural Practices) certification, which is another layer of paperwork. I asked a couple recently who ran a large goat milking operation and were in some of the large supermarkets what their annual cost was for organic, animal welfare, and GAP certifications: $200,000 a year. Dear folks, you've got to be moving a lot of product--I mean a lot of product--to justify that much expense checking off boxes and filling in paperwork.

Some farmers try to form cooperatives in order to get enough volume and clout to get into these bigger markets. Just remember that supermarkets, especially the chains, are simply the end of the industrial food system. Rather than trying to figure out how to get in them, I'm trying to figure out how to keep people out of them.

Remember, industrial commodity versus integrity craft are two completely different things. They're like oil and water. They're opposites. This book is not written for farmers who want to build Tyson chicken houses. It's written for the heretics who see a completely different path ahead. Tomorrow's opportunity in farming is not the factory outfit; it's the craft outfit.

While I'm not opposed to farm cooperatives, none has really stood the test of time. They become bureaucratic and highly regulatory. Generally, farmers have done well starting co-ops to buy collectively, but not co-ops to sell collectively. Collaborative purchasing works because everyone's interests are similar--get the cheapest price. But in selling, purposes are not nearly as similar. The in-fighting and politics siphon off much of the advantage in collective marketing.

With all of this, size seems to have an effect. Half a dozen people getting together to form a great marketing pool is not the same as a hundred. I don't want anyone walking away from this discussion saying I'm opposed to collaboration because nothing could be further from the truth. But size is a factor in anything. To assume that a small group is as effective as a large one doesn't factor in all the dynamics that a change in size creates.

Beware of empire builders. My dad used to say about people who wanted to be big, "Beware of people born with a big auger." A love affair with big has sunk many good outfits. Instead of being big, just be good. That's good enough. If big comes as a result, be thankful. If you work on being good instead of being big, you might be able to handle bigness. But if big is your goal, you probably won't be good when you get there. Just a little observation from a lifetime of experience.

If you do get into one of these big supermarket outfits, make sure it's only a fraction of your market. If it's more than 50 percent, a deal that goes south can sink your whole ship. I like to call this dancing timidly. The day the big buyer dominates your business is the day you've become a slave. I've never been able to figure out why people in the integrity farming business, who spend their lives demonizing industrial farming, think their ticket to prosperity is to dance with industrial supermarkets.

The allure of playing with the big boys is intoxicating and appeals to the basest of human ambition. But in the end it's a fool's game. With all that said, would I put chicken in Kroger if they called and wanted it? Probably. But that's them calling me; not me calling them. If and when the food scene changes enough that these big players see the handwriting on the wall and want to localize, perhaps their attitude will be more compatible. When they come courting I'm ready to talk. But by that time we might have so many alternative venues up and functioning that they'll be obsolete. That's the way of the paradigm.

2. Small, independent grocers/supermarkets.

Some farmers have had great success working with independent grocers. They are not as bureaucratic as the big boys, but they can be fickle depending on what the owner thinks.

For example, I tried to get Polyface into MOMs (My Organic Market--very clever name), a small group of grocery stores around Washington D.C. The buying agent was gung-ho on getting us in. But he reported that the owner told him, "All our customers are vegetarians. We don't want any meat." Of course, they sell meat in there, but it's from far away places. Some day, when I write my marketing book, I'll put all of these type stories in there, both positive and negative. They are instructive as to how people think. It'll make your hair curl.

We work with a small grocery in Richmond named Elwood-Thomson, located at the intersection of Elwood and Thomson streets. It's a sole proprietorship and is a wonderful full-service nutrition supplement, grocery, deli outfit with more than 100 employees. The meat department purchasing agent seems to have a high turnover and each one is a little different. What they buy and the volume fluctuates dramatically from one agent to another, but it's a great venue with lots of exposure. The beautiful thing about Elwood-Thomson's is that they celebrate seasonality and actually contract with strawberry and blueberry farmers, for example, for their entire year's crop. As the season approaches, the store uses signage and social media to create buzz as availability comes closer.

This independent grocer does a lot of community outreach. Right in the front of the store is a room with a seating capacity of about 50 where they host free lectures. I've performed there almost every year and it's a wonderful way for me as a farmer to connect with customers who buy our products in the store. These kinds of interfaces are the special things that smaller local-centric and craft-oriented outfits routinely offer as part of their persona.

Best of all, they pay faster and are far more nimble at providing what customers want. Packaging is always an issue because after all, customers see these outfits as a grocery store just like the big guys. Often customers don't know how much all that shrink-wrapped four-color perfect packaging costs. The equipment infrastructure to create that pretty packaging doesn't come cheap. As a result, these smaller stores often spend a lot of time educating their customers about wrapping and packaging.

I think these smaller outfits will be around for a long time because they can shift easier with changing buying patterns. They also imbed themselves in their neighborhoods and develop a great fan base.

One of our oldest patrons is Virginia Garden, a sole proprietorship, nearly four hours away from us in Virginia Beach. But owner Michelle Shean has been faithfully plugging away for decades. The entire store is about the size of two moderate bedrooms, but she has a loyal patron base. She communicates regularly about what's available and why. When Thanksgiving comes around she has a sign-up sheet in-store and places her order with us, comes up and gets the turkeys (or we deliver them) and everyone shows up for their turkey pick-up. It's extraordinary to see the level of care she provides in that community. If you find your Michelle, you'll be off to the races.

3. Food co-ops. These provide a perfect example of collective buying, which as I mentioned before, is where the cooperative business model thrives. Extremely sensitive to their constituency, which is also their owners, these are hard to launch but have real staying power if they survive the first few years. Just like any start-up, of course, many fail. But the ones that thrive provide a valuable service.

Just like any outfit, your ability to work with them is directly contingent upon your relationship with the personnel. Taking the manager out to lunch carries clout because these smaller outfits can make changes easier. At a big box supermarket change at the local level is virtually impossible. Corporate headquarters have multi-year contracts with non-compete clauses that lock them into certain brands and vendors. That is not so with these smaller community-organized venues.

If you're seeing a pattern here with my advice, you're seeing it right. Because co-ops have such rabid buy-in from their patrons, buzz for a tasting or in-store sampling is easy to create. Packaging is not quite as big of an issue because often shelves are stocked by volunteers. The whole presentation has a bit of a homemade feel, purposefully so. A bit ragged around the edges, the rustic nature of these venues is a good fit for craft farms.

Polyface works with two food co-ops, one in Roanoke and one in Harrisonburg, and they've both been great for our business. The Roanoke City Food Co-op actually buys whole animals to get volume wholesale pricing from us and passes that along to their patrons. In this case, the store takes the risk of moving the entire animal rather than us. They run sales and adjust prices to make slow movers go faster and fast movers slow down in order to get the whole animal gone at the same time. That's highly unusual but shows the "I'll meet you halfway" attitude of this particular organization.

I don't know how healthy these outfits will be long-term. They are under intense competition from industrial organics at the big box stores. Time will tell. Many are just hanging on; even the ones that appear to be thriving. They're a great venue and should be supported. If all the money currently going into industrial organics at Wal-Mart reverted to these kinds of venues, the prices could be lower, the food superior, and foodsheds better served.

4. Farmers' markets. Oh boy. I'm going to irritate some people on this one, but please hear me out. I'm not an enemy of farmers' markets. Some are exceptionally good and others are exceptionally bad. The problem is that in the integrity local food movement, farmers' markets have attained almost a level of cultish idolatry. To say anything negative about them is like attacking motherhood and phonetics.

Let's start with the bad stuff. Farmers' markets are primarily a social event, not a place for serious retail money to exchange hands. They're primarily upscale places where customers are there as much to walk their shampooed Fifi poodle as they are to shop. Generally folks can't buy more than a couple of things that can fit into one hand because the other hand is holding Fifi's leash.

Rather than buying a bushel of beans they buy the little baby-food-sized jar of lacto--fermented bean relish garnished with twirls of purple bow on top. If people actually did buy their week's food at Farmers' Market, the whole market would be cleaned out in half an hour.

Farmers' markets too often are viewed as destination places. Because the movers and shakers that develop them are nostalgic for the old days they often locate the market down in the armpit of the city, by the railroad tracks where the bums hang out. Now I'm all for revitalization, but I'm a farmer and I need a place where people come naturally. Like on the busiest commercial street in town, next door to Wal-Mart, for example. And most markets are only open on weekends, when people want to stay home and pad around in their jammies catching up on the news and watching games on TV. After fighting traffic for a week, who wants to fight it again on the weekend? These markets need to be open when people are already out and about.

Farmers' markets are notorious soap operas. "We can't let you sell eggs here because we already have two vendors selling eggs." "Did you see Paul's radishes over there? I don't think he grew them. We'd better report him to the market master and get him checked out." "Why is Mary here? She's from over in Boford County and that's 55 miles away. Don't we have a prohibition on people coming here from more than 50 miles away?"

Anybody who's never been in the inner sanctum of a farmers' market can't believe the behind-the-scenes drama going on. It's real theater, despite all the smiles and placid air emanating from the vendor booths.

And then you have the vicissitudes of weather. First, most

markets close for the winter. As farmers, though, we need income year round. And people need to eat year-round. This annual closing might suit veggie growers, but those of us selling eggs, dairy, poultry and meat have a real problem. Our animals don't quit in the winter. Neither do our bills. The start-stop reality is a real inefficiency in the local retail interface. Then you have the squall. If it's cold and rainy nobody comes. If it's windy everybody has to chase down their canopies before they cartwheel into the next block.

Finally, you have competition on the weekends from other social events. Any big social event, like Art-in-the-Park or Oktoberfest, pulls the crowd away and farm vendors are left to banter among themselves and buy each other's stuff.

The most consistently positive way I've seen farmers' markets used is to generate name recognition and a customer base that can gradually be leveraged into a buying club or on-farm sales. Perhaps even a CSA (Community Supported Agriculture) scheme. For a new farmer starting out, investing in a farmers' market for a season or two can get you tapped into the local food community fairly quickly. But as a long-term serious retail food interface, few farmers' markets fit the bill. My rule of thumb is that if you're not grossing in sales of $2,000 per visit on animal proteins, or $1,000 on vegetables, you'd be better off spending that time on a different market option.

All that said, here is my wish list for a great functioning farmers' market.
- Year-round.
- Inside.
- Common cash register. Farm vendors are there to schmooze and chat people up. That's part of the social element. Exchanging money with the vendor complicates everything. Can you imagine going to the supermarket and paying for each brand at the shelf where you pick it up? We'd call that inefficient. But we do that at farmers' market and call it chic. It's not chic; it's inefficient and counterproductive.
- Collaborative marketing. Producer-only needs to be moderated with reason. Most farmers are not good marketers. If three

Direct Marketing – Where

or four want to collaborate and send the one who really likes people to town with everybody's stuff, that makes marketing and environmental sense. This arbitrary and inflexible producer-only tyranny must stop. You can keep out the flea market and resellers other ways.

- Open product. Any product is welcome, even if it's already at the market.
- Vendors may sell from their entire product portfolio. Many years ago we could never make a large urban market work because we were prohibited from selling beef, since a beef vendor sat on the market board. If we could have sold beef, that market might have been worthwhile, but it wasn't due to the heavy handed protectionism (cronyism).
- Located in the commercial district. Preferably adjacent to Wal-Mart.
- Non-weekend hours. Perhaps Monday and Thursday 3-9 p.m. Catch people when they're already out and about.
- On site entertainment. Ratchet up the theater shtick by always having food to buy and eat on-site. Make sure local magicians, musicians, jugglers, balloon artists--you get the idea--are always there to entertain and excite the kids.

That's it. Any farmers' market that implements these ideas would see their sales triple, I'm sure of it. In case anyone thinks these are too hard to do, I've seen all of these at various farmers' markets around the world. But I don't think I've ever seen all of them at one. If one would put them all together, Katie bar the door. And if it did, I'd want to be there.

5. Electronic aggregators. Now we're getting into the fun stuff. Sometimes called virtual farmers' markets, these outfits are springing up all over the country and represent the cutting edge of retail. Of course, Amazon is the big gorilla in this space and everyone localizing in the food space is using software and protocols designed by Amazon.

The internet is becoming more ubiquitous by the day. I'm 60 years old and still can scarcely fathom what it has done to our lives in such a short space. It must have been like what buggy makers felt when the Model T came out. Only at warp speed.

The fact is that the retail brick and mortar physical retail interface is expensive to maintain. Public bathrooms, handicapped parking, proper lighting and spacing so nobody will sue you when they trip and fall, cashiers, accessible shelving. All of this adds to the overhead.

But if you can aggregate electronically and make the transaction in a non-public, portable space, suddenly all the overheads shift. Now the retailer does not need to maintain an expensive facade; instead, he can just occupy space in the Cloud. In classic business terms, this is a major disruption. Not business as usual. Suddenly everything we thought was important becomes obsolete. It's no longer important to have someone spruce up the bathrooms every hour. It's no longer important to position the right products on end caps. It's no longer important to know which shelf gets which products.

Without a public physical interface, the retail presence simply occupies the screen on our computer. Taking this idea and running with it, entrepreneurs developing this idea are fast carving out identities in the food space. When you couple this with the Uberization of everything, it's quite a combination. These outfits don't hire delivery people. They use independent contractors. No employees, no paperwork, no responsibilities, lower overheads.

Customers order from a virtual shopping cart from farms in their area. Often the number of farms is quite large, even numbering a hundred or more. The virtual shopping cart preserves brand integrity so customers can shop from their favorite farmers. The aggregator maintains a minimal inventory at a private warehouse. That warehouse is stripped down and easy to maintain--no public restrooms, no high lumen lighting, no polished floors, no cash registers, no public access liability.

A robot receives a purchase order and goes down the shelves, assembling the customer's box. The box arrives properly packed and sitting on the dock ready for delivery. The delivery trucks are equipped with special GPS software. The driver plugs in his destinations for the day and the software automatically programs a route to eliminate left-hand turns. That alone drops 15-20 percent from the fuel bill and 20 percent from the labor/time bill. In Great Britain, it would eliminate right-hand turns.

Customers rendezvous with the truck at pre-determined pick-up spots. Some make door-to-door deliveries. The point is that you can buy from your home and nobody has to fight traffic and sit in the left hand turn lane waiting to get into the supermarket. Some of these outfits are adding toilet paper and other non-food items to truly create the virtual supermarket. Anyone watching retail marketing trends has to admit that this model is going to grow. It's gaining steam and it will not go away.

How much it will disrupt the brick and mortar interface is anyone's guess. Of course, Amazon, in this space, handles regular industrial food alongside organics. That is a nightmare for big box physical retailers.

Plenty of mail order and smaller versions of internet sales are entering this space. They're all permutations on the e-commerce theme. Tomorrow's successful farmer will either need to hook up with one of these or develop an e-commerce sector in order to be successful. I think the demand and the social media component of these systems will offer farmers additional marketing opportunities without the normal condescension seen in the big box store phase. Because this marketing model is coupled closely with the transparency of social media, I think the anti-farmer creep that permeated the big box supermarket system will not be as acute. Time will tell.

Social media changes the conversation. One of the pillars of the industrial food system is opaqueness. The "No Trespassing" signs erected around factory farms are symptomatic of a secretive system. Social media, with its open sourcing and viral gossip mentality, is the opposite. Some people who watch these trends think that the democratized conversation of the internet has rendered industrial regulatory oversight obsolete.

Back when the big food companies developed, the industrial economy reigned supreme. In perfect harmony with physics, where for every action there is an equal and opposite reaction, this industrial power that catapulted the US into dominant world manufacturing position ultimately disempowered the individual. More and more Americans grew up feeling like just a cog in a wheel. This spawned the whole counter-culture revolution of the 1970s, "the beaded, bearded, braless revolution" as one historian said.

This individual disempowerment in the face of cultural muscle moved people to ask why and not just how. Part of the backlash manifested itself in La Leche League and health food stores. Who would have thought, in the 1950s with breast feeding banished to barbarism and progress wrapped up in Infamil and Similac, that within a generation we'd be nursing babies again and within one more generation we'd be linking breast feeding to a reduction in breast cancer? Who would have thought?

Since literature, poetry and art tend to always be on the cutting edge of societal evolution, it's fitting that as we enter this e-commerce era, media would lead the field. But other things, from automobiles to food, will come lumbering along. Media will develop the software, the transaction protocols, the distribution networks. That will pave the way for everything else to follow.

The level of interface that farmers can have with their customers, through these electronic aggregators, is tremendous. Virtual tours of the farm, direct interaction with the farmer through social media. Who would have considered that when spraying green beans on a Jolly Green Giant farm? But with information being democratized and opaqueness giving way to open sourcing, secrecy giving way to conversations, the farmers' access to the consumer flattens. Innovation in this space is happening so fast right now I hate to even write about it, knowing that within two years even this discussion may seem archaic.

A savvy social media farm can nose into these spaces far cheaper and easier than trying to get on the shelf at Wal-Mart. The farmers who occupy these virtual shelves on these local-centric electronic aggregators today will be the ones who create loyal market share and brand identity for tomorrow. This is where the war is being fought today and like it or not, farmers who want to be successful will have to participate.

Perhaps the biggest retail battle brewing is between Wal-Mart and Amazon. It's a heavy-weight fight, for sure. Wal-Mart is offering more and more electronic service--who wants to stand in a cash register line? And Amazon is looking at physical presence in key markets. What an amazing war to watch.

6. Restaurants. As my daughter-in-law Sheri says, chefs are an unusual combination of artist and organizer. Normally when we think of artists we think of spontaneous whatever kind of folks. But chefs are not whatever kind of folks. They have to keep a bunch of responsibilities on time and on target. They're time oriented. All their artwork needs to be on schedule and customer-perfect. They live in a pressure cooker of time, motion, and creativity.

Chefs have tight schedules. Never call on a chef during her busy time. Here is where some website searching can put you ahead of the game. Obviously if the restaurant serves lunch, don't call at noon. If it serves dinner, make sure you call before 4 p.m. Use the website to peruse the menu. Is the restaurant a fit for you?

Generally speaking a restaurant price is divided into three fairly even parts. First is ingredient. Second, labor. Third, overhead. So if you see an entree for $9.99, you know that your grass-finished T-bone at $10 is not going to work. Do your sleuthing and figure out which restaurants are compatible with your farm. Chances are you won't be going to the cheap places.

Once you've picked a fit, you need to get a sample to the chef. You never want to just take a sample by the restaurant because the staff will pounce on it like buzzards and take it home. You have to make contact with the chef, which can be easier said than done. These are busy, high performance folks. Be persistent. Sheri hounded one chef for six months--yes, called him every week for six months--before he finally came to the phone and talked with her. We gave him a sample; he became a customer; wonderful story. But don't miss the back story--a persistent knocking. Don't give up unless the chef tells you not to call again. "I'm busy" is definitely not a "no."

Their restaurant is all about them, not about you. Stroke the chef. Tell her she's amazing; ask how you can help her achieve her goals. Listen. Sometimes a size, portion, or packaging issue stands in the way of a big sale.

Restaurants use a lot of stuff, consistently. They can give you quite a bit of exposure in the food community. That's all the upside. The downside is that they are heavily regulated with food police. Be careful. And chefs can be finicky.

You will find chefs too difficult to work with. Some will want you to do things for them that you can't do at a competitive price. For us, that's offering boneless skinless chicken breasts. We're not automated here, so this is a laborious and tedious luxury. Homeowners will pay for it, but not chefs. Explain your situation or limitation--chefs understand margins and efficiency.

If the chef wants tiny carrots, charge accordingly. You can't handle a hundred tiny carrots for the same price as 50 large carrots. Sticky situations like this need clarification. Often chefs are as excited to learn about your farming world as you are about their culinary world. Give them freebies from time to time. When we have a new batch of pullets beginning to lay, we often take in a case of smalls and give them to the staff. That's more valuable for a $50,000 a year client than selling them for 50 cents a dozen.

Go eat at your restaurants. Make sure the chef knows you're coming. Restaurants that feature local farmers go nuts when their farmers come to dinner. Participate in "meet the farmer" events and build relationships with your chefs. One of our most memorable chef interactions was with our very first chef. We invited her and her husband out to the farm and went up into the woods, built a fire, and cooked hot dogs on sticks. Then we graduated to s'mores. They've since moved out of the country, but they still remember that evening fondly, as do we. You don't have to do something fancy; give the gift of time and you'll endear yourself to them for a lifetime.

When you get an account, be punctual. Remember, these chefs are schedule oriented. Call at the same time every week; deliver at the same time every week. Think about your own kitchen. It's a routine place in a hectic world, right? Magnify that by tenfold and you begin to grasp how it is in a restaurant kitchen. Chefs love dependability. If you say you'll do something or show up with something at a certain time, do it. Their biggest chore is finding dependable people who show up on time. Most chefs feel like glorified babysitters, trying to pacify the front of the house and the back of the house. They ought to carry around pacifiers so they can stick them in the appropriate mouth. You be their shining star.

7. Community Supported Agriculture (CSA).

This model had just leaped across the Atlantic from Europe when I was getting started in farming at the beginning of the 1980s. It has certainly stood the test of time and is undoubtedly here to stay. In my heart I love this model, but it has a lot of built-in hurdles. And remember, I'm all about reducing hurdles. The problem is that you have to buy the whole box.

In case you're uninitiated on this concept, it revolves around buying a share of the farm's produce. Rather than buying a carrot or a chicken, you buy a share of the aggregate production pool and the farmer puts into that share an equitable portion. The permutations on this theme are astoundingly innovative. You can buy beef shares, for example, in which the farmer packs 1/40th of an animal. You can buy vegetable shares (the most common) in which the farmer might pack as many as 30 varieties of produce.

The beauty of this system is two fold. First, the courts have ruled that this arrangement is not a sale. That means the transaction is not "in commerce," which is the legal phrase used to denote government regulatory oversight (what I call action by the food police). Where raw milk is illegal to sell, for example, farmers are using the herdshare loophole, whereby you purchase a share in the herd and that entitles you to a portion of the herd's milk every week. From a liability and regulatory standpoint, therefore, the investment-dividend concept, or the owner-caretaker relationship, offers some real benefits. Of course, since it's not a sale, you don't have to charge sales tax. That's cool.

The second beauty is that the farmer gets money up front. In typical produce CSAs, the most common type, shareholders buy a share, or at least a portion of it, up front and the farmer then uses that money to finance seed, fertilizer, and other purchases to maintain a positive cash flow. If a share costs $250 for a season, for example, the shareholder may pay $125 up front and then the rest midway through the season. A hundred shareholders therefore would create $12,500 for the farmer to start the season rather than waiting until the first harvest to begin pecking away at the up-front expenses. That's a beautiful arrangement, but it takes die-hard local food aficionados to play the CSA game. That's a small subset of the food-buying public.

The sad truth is that CSAs have a high failure rate because only the most gifted farmers figure out how to offer the proper combination in the box. Vegetables are easier than meat, but it's still difficult. As I travel around, I'm seeing more and more hybrid CSAs. They still do the box, but always have an a-la-carte option on site. This way people can augment their box with personal choice.

Die-hard CSA farmers rant and rave about how people need to learn to eat a wider variety of stuff, and creating that box insures that patrons (subscribers or shareholders, as they're called) experiment with new things. That may be true, but the reality is that most of those odd things go out in the garbage. Remember, we like routine. We do. That includes food as well as it does our favorite music. Why are favorites what we like to hear? Because they're familiar. New things are risky.

Perhaps the hardest part of a CSA is weathering a debacle. The flip side of the CSA is that since subscribers actually pre-invest in the farm's production, they do not get their money back in a crop failure. If the fox wipes out the chickens, too bad. If a flood wipes out the carrots, too bad. So while the concept on its face sounds stable and insular for the farmer, the need to perform is there nonetheless, in some ways more than in a-la-carte systems. At least if nobody has paid for anything you just tell them you don't have it. Although CSA farmers spend a lot of time dotting i's and crossing t's in order to make sure their subscribers understand that this is not a sale, at the end of the day, subscribers want something for their money. It's an unusual person who will swallow getting nothing for her investment.

Although here at Polyface we have never done a CSA, I've seen too many successful ones to discount them. Given the right gifts, the right community, the right narrative, I think CSA is a viable and beautiful marketing option. In a couple of years we might have one here; you just never know. It sure solves a lot of regulatory, liability, and tax issues. As regulatory, insurance, and compliance burdens escalate, this option may look more attractive in the future.

8. On-farm sales/store. Even farmers not gifted at marketing can sell something off their back porch. My completely intuitive, non-scientific assessment of what almost anyone can sell from their farm house is as follows:

- 30 dozen eggs a week
- 300 broiler chickens a year
- 5 beef animals a year
- 12 hogs per year
- 20 turkeys at Thanksgiving
- 10 families buying produce

Anyone doing this much business is not really in business. But as soon as you go above these numbers, everything changes.

Permutations on this theme abound, from elaborate 24/7/365 farm stores on busy highways to a garage corner with an honor box in it. The advantage of course is that neither the farmer nor the food has to travel. It gets customers out to the farmer, where the ultimate loyalty and connection can be cultivated to perfection.

The disadvantage is twofold. First, it requires the customer to come out to the farm rather than some more central place. Location is obviously a big variable here. If your farm is on a major highway, this is probably not a problem. But if you're 100 miles from a Coke machine, on-farm sales can be problematic. Our farm is only 8 miles from Staunton, a town of 20,000, but we're on a dirt road in the middle of nowhere. You'd be surprised how many people refuse to put their car on a dirt road.

Second, it brings people into your space. That's a potential problem for both introverts and extroverts. For extroverts, it means you'll have to fight the temptation to yak with everybody. Trust me, you'll get to the end of the day and realize you haven't accomplished anything except kibitzing. It's a problem for introverts who feel like people are invading their space. I'm a big believer in on-farm stores, but they should be disconnected from the house and have clear open and closed times.

The propensity to drive to a retail location is in direct proportion to the value of the purchase. People don't bat an eye driving an hour if they're going to spend $500. But they balk at driving 5 miles if they're only going to purchase $20 worth. Anything you can do to enhance the on-farm commute will help overcome this reluctance. Make sure you have clear signage so folks who come know where they can--and can't--park. Make them feel welcome with a nice big farm sign. Clean, gravel parking lot. A store welcome sign properly decorated--some flowers, sitting benches, picnic tables, sand box for kids to play in.

All these things enhance the farm experience. Play to the children. If the children want to come back, Mom and Dad will come back. Have toys in a corner, perhaps some wooden blocks, Lincoln logs, carved farm animals and toy farm equipment or Tonka trucks. A designated play area where the children's stuff can stay put and not migrate all over the farm will help keep this from being a frustration.

Marked trailers, a map, and even explanatory placards around the farm present a professional look and add to the edu-tainment value. Most zoning ordinances allow you to sell a few things from the farm store that are not your own (in our county, it's $100,000 a year without a license). Take advantage of that to stock items that will compliment your own. A little of this goes a long way and gives people more things to peruse and buy.

In the store, clear signage and pricing is especially necessary if you have an honor box. If you use that system, be sure to have a couple of calculators, invoices, scratch paper, and ink pens handy. I'm sure some people are aghast that I'm suggesting the honor box as a viable option, but I've been on many farms and have never heard of theft. The secret is to have a big enough or heavy enough box, and have it locked, with a small enough port hole, that nobody can walk off with the money. People don't steal food; they steal money.

Use your social media to drive traffic to the store. Run sales, specials, events, and anything else you can think of to generate interest in folks coming to the farm. We allow people to wander anywhere anytime unannounced. That open door policy engenders tremendous trust and public relations. It's a show stopper for differentiation because what self-respecting business allows people who may not even be customers to wander willy-nilly backstage and see anything they want? That's our credibility and we've staked our reputation on it.

9. Metropolitan Buying Clubs (MBC).

This is a serendipitous invention we developed in the early 2000s. We had four ladies who would come down once a quarter from eastern Maryland, about four hours away. They'd call the night before to let us know they were coming.

It was a ladies' day out for them. They'd leave home at 7 a.m., hit a couple of antique stores on the way down the valley to see if there was anything left that the Yankees hadn't carted off yet, have lunch in Staunton, and then arrive at the farm around 1 p.m. They brought an SUV full of coolers and really bought stuff. In fact, they would buy around $800 per trip. I kind of liked these Maryland ladies (that's what I called them).

This went on for nearly two years until one of them pulled me aside and said:, "You know, we have a lot of friends."

I said quickly, "I think I'd like to meet your friends."

She continued, "We've tried to get them to come with us. We've offered to car pool, to pick stuff up for them, but they say they're too busy with soccer, ballet, church meetings, whatever. I've been thinking. What would it take for you to come to us with a delivery?"

In typical entrepreneurial gut instinct fashion, I cogitated momentarily and replied, "$3,000. You get $3,000 of sales together and I'll come to you." I didn't want to deliver. I wanted to stay on the farm. I'd already been able to subcontract out the restaurant delivery deal, so I wasn't ready to get back into pounding the pavement routine again. I figured that number would stop them.

What I did NOT know was that one of those ladies was an urban metaphysical guru. She'd written six books on how to get in touch with your inner being, know thyself, release the power that is within you. The next week at class she announced to her disciples that she'd found a clean food option and "this sheet is where you sign up." By the end of class, she had $3,000.

She called me the next day and asked me when I was coming. Dear folks, what do you do with a phone call like that? DRIVE\ baby! That's the way this marketing venue started. It uses internet shopping cart software with sophisticated communications software (to get through spam blockers) coupled with a Quickbooks integrator to turn the orders into purchase orders. I take no credit for all this--Sheri has done it. Her software cheat sheet is on our webpage at *polyfacefarms.com* under the RESOURCES tab.

Many, many farmers around the world have adopted this idea with great success. The beauty is that it allows an extremely rural farm like ours to access the urban market with non-speculative sales (everything on the truck is pre-ordered) with a volume that creates economies of scale in the distribution. It allows customers complete choice from our inventory and enables them to shop at their convenience instead of during open store hours or at some other location. They can lie in bed at midnight with their significant other and wrangle over porterhouse or tenderloin without taking up our time in front of the cash register or farmers' market stand.

We prefer private homes for the drop sites to preserve the mystique of personal attention and down-home ambiance. Hostesses receive a 10 percent discount. We serve each drop eight times per year at roughly six week intervals. All meat is frozen. Each person orders on a rotation when the time is open for their drop; they can change the order any time during that week up until it closes, which is two days prior to the drop. We put each order in separate coolers. If it's small, we put it in a cooler bag and can place several of those in a cooler. We have about 200 coolers--they're a lot cheaper than a Thermo-king refrigerated truck.

We try to service three drops in a day, getting in after rush hour and getting out of the city before rush hour. The drops run from about 10 a.m. to 3 p.m. Everyone meets us at the drop, finds their own invoice (alphabetical order in a stack on a table), and pays. The invoice designates what cooler or cooler bag contains their order. They put it in a bag or box they brought and away they go. Often customers are only on site for five minutes. It's extremely efficient for all of us. We can move $5,000-$10,000 worth of product in less than half an hour with only one staging person.

The only problems we've had are a couple of nasty neighbors or homeowners' association rules. Isn't it interesting that an HOA has no problem with the UPS or FedEx truck stopping at every house, but if all those people come over to one pickup point, they throw the delivery bum out. Insane. And I'll bet many of them support the Nature Conservancy. Aaaaargh! Obviously we have to be able to get off the street and be in a spot where our patrons are not a traffic hazard. We don't sell on site. We take no a-la-carte stuff. Nobody can walk up and buy anything. We're just delivering orders like the FedEx guy.

It's a fast turnaround. We're in and out quickly. We can take things from other farmers. We can even take lumber or crafts. Since it's our deal, we're not bound by farmers' markets rules and regulations. It's not a CSA because folks can cherry pick freely, from half a dozen eggs to a whole beef. We have not used this concept with produce yet, or something highly perishable like milk, but we know people who are using it for dairy and it works well. I think it would work for produce as well but you would have to go weekly. That would mean starting with a smaller and cheaper delivery vehicle and then scaling up as volume escalated.

The MBC now accounts for 35 percent of our sales. New customers have a place to give us the name of the person who recommended them to us so we can give $10 credits to those cheerleaders. With these $10 credits per new customer, some people run this like a multi-level and are spending less on our food than if they bought junk at the supermarket. For us it seems to take the best of all these other venues without the negatives. If someone doesn't show up, they pay a restock fee. If we mess up their order, we give them the restock credit. All in all, it allows us to penetrate farther into the food-buying pool even though we don't live next to an urban area.

10. Food trucks, kitchens. I can't help myself. I want a food truck. Why the sudden American craze over food trucks? Did Americans suddenly wake up and want food from a chassis? What's going on?

The simple truth is that food trucks are a way in for culinary entrepreneurs who don't have enough time, money, or emotional savvy to navigate the expensive and troublesome labyrinth of regulations to open a brick and mortar facade. You don't have to worry about on-site insurance, a parking lot, bathrooms. It's a way in and simply reflects how onerous the bureaucracy is these days. Our town of 20,000, Staunton, Va., had a "Battle of the Food Trucks" last year and attracted 36! They divided them into four categories, eight trucks apiece. It was insane.

When you think about what farmers spend on things, $30,000 for a tractor here and $20,000 for a baler there, suddenly a food truck doesn't seem too far off the reservation. Ditto a kitchen to value add your cabbage into sauerkraut and your cucumbers into pickles. If you create a market brand for relish and pickles, how much land do you need to be farmer? Not much. An acre is plenty.

I understand that a food truck or kitchen is expensive, but in the big scheme of things, they're often the best investment a small farmer can make in order to value add limited products to a full-time living. Just like marketing, this is another aspect conducive to partnering. If you have no interest in culinary entrepreneurship, that's fine. Find a partner who does. Create something together and you'll lock in your market through the value added product. Now you have a salvage for perishables and a way to insure year-round cash flow.

I know this has been a long chapter, but I hope by now you realize moving your farm products is as important as growing them. Fortunately, we have a lot of marketing options today. Chances are you'll experiment with a couple of these venues and settle on the ones that work for you. That's as it should be. Here at Polyface, we don't use all of these. But somewhere in this list is a fit for you, and when you develop it, you'll enjoy the satisfaction of middleman profits.

Chapter 8

Direct Marketing – How

Marketing precedes sales. Lots of times we use the two terms interchangeably, but actually they are different. Marketing is the overall strategy or campaign to enable sales to happen. Sales are the actual cha-ching of a transaction.

Marketing is all about messaging, position, and technique. Sales is converting that into a yes on the part of a buyer.

Here are what I call cornerstones of farm direct marketing.

1. Diversified portfolio. Anyone who does financial investment consulting will encourage clients to buy a lot of different things. You want some high risk and low risk, some retail and some research. The idea is that if you spread your investments, averages protect you from wild swings in one sector or another. Rather than throwing all your balls in an aggressive basket or a conservative basket, you hit a happy medium and ultimately that buys some stability.

The same thing is true with marketing from the farm. One of the attractions of the supermarket is that you can do so many things with one stop. I'm absolutely convinced that one of the reasons Polyface has been successful is because almost from day one we offered a range of products. We created a one-stop shop for meat, poultry and eggs. Folks could get beef, pork, chicken, rabbit, turkey and eggs. If it had been legal, we would have offered milk. The day it's legal, we'll add it to our portfolio, but Virginia is a commonwealth of tyranny on that regard. Don't get me started.

Nobody makes money on a first time sale to a first time customer. Acquiring customers is costly, and it always eats up the margin on a first time sale. The cost of maintaining loyal customers is far less than the cost of attracting that customer in the first place.

As electronic interfaces chip away at brick and mortar stores, supermarkets are adding more and more items to get people in the door. Who would have thought just a few years ago that you would go to the grocery store to do your banking, get a manicure and haircut, take your yoga class, all while the kids are being entertained by story time? You can get your oil changed, buy diapers, and pick up bananas all at the same place.

As direct market farmers, this is our competition. We certainly don't want to duplicate a supermarket, but we need to appreciate what attracts people there and utilize what fits. One of the things that fits is the diversified product choice. A customer who buys beef wants to buy pork. A loyal customer always looks for more ways to patronize a business she believes in.

In retailing, we call this bounce back. We're never looking at a sale as an end, but a beginning. We're always thinking about how to get that person back in the store, to look at our website, again and again and again.

The axiom is this: it's much easier to find 100 customers who will spend $1,000 with you than it is to find 1,000 customers who will spend $100 with you. In either case, the total sales is $100,000, but dealing with 100 customers is far more efficient than dealing with 1,000 to get that $100,000. If people are going to darken your venue, whether it's at a farmers' market or an online shopping cart, once they've made that investment to be there they want to feel like their investment in darkening your venue to shop was a worthwhile use of time and effort.

The more they can buy from that one stop, the more efficient they'll feel about the purchase. Making customers feel efficient and rewarded is part of building loyalty. This doesn't mean you have to personally produce everything you market; it does mean that the larger your portfolio the easier it is to attract buyers. The person who makes the sale owns the customer.

If you position yourself as the retail interface for your clean food farming community, you'll be the leader in that space. Once you have a happy customer, she's always looking for the next thing. Just like the cut up chicken. We already had a customer base. The customer base is always asking for what's next. The opportunities for products is literally limitless. Look around at other farmers in your community; what else could you offer? Don't forget about non-food items like homemade wooden children's toys or crafts like seasonal Christmas wreaths.

If all you offer is beef or bison or lamb, it's hard to keep people on board. One of the reasons we collaborate with other growers in the community to service our restaurant chefs is because it positions us as the defacto go-to place for everything on our delivery truck. Even though we preserve the vegetable growers' identity by stapling their invoices under our Polyface aggregated cover invoice, the chefs know it's coming on our truck and their check is going to Polyface. You don't have to handle the product very much at the point of sale to acquire a kind of owner perception.

"If you want this, you have to come through me" is a wonderful marketing position. Part of branding and marketing is to make the customer dependent on us, that we're indispensable. That's every marketer's dream.

If you're coming to this farming gig with one production item, look around at what you can add. Vet the other items and be the conduit in your clean food community. Chances are the other farmers in the area are desperate for someone like you to take over some of their marketing burden. Look at yourself as a catalyst, a facilitator. So enable your customers to spend more and more of their shopping dollars with you by diversifying your portfolio.

2. Differentiation. It's a noisy world out there. Everyone selling a service, idea, or product jockeys for a buyer's ear. Look at junk mail and robo callers. Every year we think we can't be assaulted any more aggressively and then there's a new angle. Just when we thought the internet would free us from advertising, it's become the new conduit and you can hardly pull up anything without having to close down an ad.

I remember well the first time I heard Michael Olsen, author of *Metro-farm* and guru of farm direct marketing speak at a conference in California. It was in the late 1990s, in California, to an audience of sustainable farmers. Attendees were by and large earth muffin, tree hugger, liberal Democrat, dread lock hippie types. Michael stood up and bellowed: "Marketing is WAR!" You'd have thought someone just got shot.

The whole room was in shock. For effect, he bellowed again: "Marketing is WAR!" Folks squirmed. They didn't even believe in the Pentagon, let alone war in the garden. I'll never forget my satisfaction watching Michael challenge these farmers with their new reality. The "grow it and they will come" idea is simply not true. People who think I have some sort of magical karma that makes people want to buy from me have no clue how much effort I've put into creating and telling our story.

People need a narrative in order to buy from you. That means you have to talk yourself up, compare yourself with the competition, and break through their noise to get their attention. "Hey! You! I'm over here! And I'm different--so much so that if you'll listen to my pitch you're gonna love me."

Isn't every ad about differentiation? Ours is prettier. You'll be healthier with our drugs than those drugs. We have a better price. The art of selling is the art of winning people to your side. You want to capture them, make them surrender to your product. Lay down their arms, pull out their wallet, and let you be in charge of their life. Okay, maybe that's a bit too far, but folks, marketing is about creating a narrative people love.

That means your product has to be as good as described. Few things irritate people more than being promised something better and then finding out it's the same stuff everyone else sells. Today, with social media, this bad feeling isn't limited to people within earshot. It often gets played for the whole universe. People who feel played or betrayed now have social media to exact revenge. It's a potent field leveler.

From day one, I created a slide program (yes, on a Kodak carousel, remember those?) and presented it to civic clubs in the area. It was not an overt marketing plan. It was about how our farm was healing the landscape, and how animals could build soil, purify water, and clean the air. People loved that narrative. It was something they could get behind and actually embrace.

To be sure, most of the narrative in the food industry is "ours is cheaper." That's certainly a story of differentiation. "Buy here, save more." "Pile it high and sell it cheap." "Home of the $1 menu, now for a limited time." The dominant theme of the American food scene for the last couple of generations is about price. As the populace becomes more aware that this price woefully fails to capture externalized costs, a new narrative is emerging.

Never before has it been easier to attract attention by saying things like "we build soil." "We sequester carbon." "We increase biomass production to inhale carbon and exhale oxygen." Certainly I'm not asking you to make a claim that's untrue; I'm merely suggesting that the cultural awareness of industrial evils creates a space--a big space--to offer an exciting new narrative.

Your product must be epiphanal. It needs a WOW! factor. This is the whole idea behind Seth Godin's *PURPLE COW*. Anyone can have a red cow or a black and white cow. But who wouldn't go out of their way to go and see a purple cow? As a marketer, you need to figure out the core of your story. It may come from your mission statement or your vision. Your story needs to be consistent, clear, and concise.

Some farmers have chosen the heritage breed mantra as the core of their story. Others have chosen nutrient density. Others local transparency. The cool thing about living today is that the number of options has never been greater. As industrial food becomes worse and worse, the options on the other side open up.

If you're a bit lost in the philosophy behind this statement, it's explained quite well in *YOU CAN FARM*. I've purposely stayed away from material in that book while essentially building a graduate level course in this one. I know a lot more now than I did then, but if you're new to all this, go back and take a look at the elementary level.

Remember, the more things you try to be, or claims you try to make, the more difficult the narrative. If you clog it with too many things, it gets long and laborious, eventually losing the interest of people you're trying to reach. Keep it simple. Eliminate prepositions. Use active voice, not passive. Don't say "we have been building soil." Say "we build soil." Don't say "we have been loving our animals for five years." Say "we love our animals." Don't say "we've been planning and planting our garden for you." Say "we plant new vegetables every day just for you." See the difference? If that's confusing, ask an English teacher.

Make it punchy. You should, like, you know, try to not use as many words and, like, say awesome and meaningful things, right? And yes, take your brochure to a good writer in the community, maybe an English teacher. Sweating the details is itself different than what everyone else is doing. You'll never regret investing the attention and time in your narrative to make it Goldilocks perfect: "Juuuuust right." It's representing your business, so invest what's necessary to do it and say it right.

What's the most distinctive thing about your farm and about your products? Brainstorm individual words or phrases. Pick the most poignant ones. Those provide the core of your narrative. Struggle with it. Work with it. Rearrange them. Different combinations will bring new perspective and new ideas. Try them on the battlefield; listen for responses. Which ones resonate and which ones repulse or just leave folks blah? As you probe the marketplace, hone the message. Be positive, but be precise. Don't be afraid to challenge.

For decades I've referred to factory farmed chicken as fecal soup. It catches people off guard, to be sure, but it sure gets the point across. It's so far out in left field people don't have time to dislike it. It's a home run phrase. I always say it with a laugh to juxtapose the heinous metaphor with humor. Contrast works, like sweet and sour, soft and crunchy. Remember, marketing is war, and therefore not to be taken lightly.

3. Customer friendly. Part of marketing has to do with breaking down hurdles. Is your marketing plan easy to find and easy to activate? Would you buy from you? This can be everything from the actual sales transaction to the venue to the packaging to the price.

Prices need to be clear and visible. People don't like confusion or things that appear untrustworthy. Being upfront and precise about pricing is not only professional but critical to establish trust. The shyster always hems and haws around. If you have three different prices for something, lay it out clearly. If you have wholesale and retail prices, create clear price sheets so customers know exactly what they're getting for what price.

In the end, while this may sound time-consuming, it's actually time-saving because it saves you from having to remember what you told client A so client B doesn't compare notes and feel like you're treating buyers unfairly. Few things destroy trust faster than a sense of unfair treatment. A comprehensive and clear price list is one of the best ways to dispel fears of hucksterism.

Is it by the pound, by the bunch, by the package? A good price list isn't easy to create, but it will offer peace of mind both to you and to your customer. Always put a date on it and a "price may change at any time" caveat so if somebody picks up an old price list, you have an out if it has changed. The idea is for both you and a potential buyer to examine efficiently every option you have. People can't buy what they can't see.

Our price lists always include another caveat at the bottom: "Want something you don't see? Ask. Maybe we can do it for you." This appeals to the most exotic customers and encourages feedback.

How do you know what products to add if you don't encourage customers to tell you what they want?

I think here at Polyface we were one of the last to join the electronic transaction revolution, but we finally succumbed to the crush of requests to take credit cards. Anything you can do to make it easier for your customers to buy is a good thing. If you're dealing with a lot of cash, try to eliminate change and pennies. If possible, price so sales tax is included and keeps things at even money.

Customer friendly venues are critical. One nuance of customer friendliness is the tension between simplicity and choice. In *REWORK*, Jason Fried and David Heinemeier Hansson explain that, "the menus at failing restaurants offer too many dishes. The owners think making every dish under the sun will broaden the appeal of the restaurant. Instead, it makes for crappy food (and creates inventory headaches)."

Sometimes customers can become paralyzed with too much choice. This came home to me when we hired our distribution manager, Richard. He lived in northern Virginia and had been a customer for some time. When he saw our request for this need, he jumped in and saved Polyface. At least, that's how we feel. He's a Weston A. Price Foundation (WAPF) follower, which means he's all about heritage foods and domestic culinary arts.

He's not scared about fermenting crocks percolating in the corner of the kitchen. Cutting up a chicken does not daunt him. He's savvy about food and has literally redirected his life trajectory by going from morbidly obese (his book, *A LIFE UNBURDENED*, is a must read) to running marathons. What I'm trying to get you to understand is that he has drunk our Kool-aid. He's all in.

So imagine my surprise one day during a conversation when he sheepishly admitted that it took him three years to get up enough courage to buy half a hog from us. "Richard, what on earth was scary about buying half a hog?" I asked incredulously.

"Well, I'd never done it before. And I had to call the butcher to tell him how to cut it up. I'd never talked to a butcher before.

Those guys wear white aprons with blood on them. They have cleavers dangling from their belts. I didn't know what pieces came from where. The whole process was intimidating and scared me."

Wow. Here at Polyface we thought we were offering this wonderful choice to our customers. You could buy a pork chop or a pound of sausage, but if you wanted to buy in bulk and get complete control over what you got, you could buy half a hog and have it custom-cut to your specifications. What could be more customer friendly than that? But from Richard's perspective, it threw a whole set of intimidating hurdles in front of the customer.

Then he told us a story about his Information Technology days, when he'd been a software designer for a hot mid-Atlantic-based company. This outfit would build systems for clients and their whole strategy was customized software. One day someone had the brilliant idea of creating four packages and eliminating the customization option. Immediately sales quadrupled and profits went off the charts. The problem was too much choice, and it paralyzed would-be customers.

Armed with Richard's encouragement and stories, we decided to move our customized volume sales from hanging weights with custom-cutting to several pre-cut options. Beef quarters became five options: All American, Slow Cooker's Dream, Country Club, Deluxe Delight, and Meatloaf Mania (all ground). Sales literally tripled overnight. We couldn't believe it. With these options, customers who wanted the volume discount offered on a whole quarter no longer had to talk to the folks at the abattoir.

If you've ever noticed, most abattoir operators do not exude the most gracious people skills. They're no-nonsense folks and often a bit gruff. They frankly get tired of dealing with newbie urban folks who ask, "what can I get from a loin? Where is it? What do the pieces look like?" So the frustration runs both ways. This simpler option allowed us to preserve--and encourage--volume buying by making it simpler. We took out choice and abattoir interface steps, both of which yielded a much more streamlined transaction.

For pork, we offered four options for halves: Chop Lick'n, Loin Party, Saus-pan, and Pork Lickin' Good. Each is cut and packaged a bit differently, but each contains a half a hog. One of the biggest positives this created for us was that we could now sell the volume for its actual retail weight rather than hanging weight. Customers routinely accused us of stealing their meat.

Selling by hanging weight, which has been an industry standard since the beginning of scales, is the skinned or scalded product prior to cutting. During the cutting process the butcher discards discolored pieces, some gristle, and of course lots of bones. I wish I had a nickel for every time someone called after picking up their quarter or half, "Where's the rest of my meat? The invoice says 264 pounds but I took the boxes into the bathroom scale and it's only 215. Where's the rest?"

A ten-minute discussion later, the customer reluctantly understood about the trim and bones. Now we no longer have this conversation, which has greatly simplified our lives. The point is that bundling and packaging can be key to simplifying the transaction in order to make it more customer friendly. And yes, we still do offer the hanging weight option for the most savvy of our customers, but they are knowledgeable enough about the process to enjoy calling the butcher.

Sometimes less is more. Simplify, simplify, simplify.

4. Delivery is a separate business. Most of the great ideas I've had have been serendipitous. I absolutely stumbled onto this one. When we began delivering to restaurants in the mid-90s (that's 1990s, not 1890s, you young whipper snapper wise guys) I knew I didn't want to do that drive forever. The whole goal was to start it and then let someone do the delivery as a subcontractor.

That pushed me toward separating the delivery from the product, on the invoice. In other words, we'd create a subtotal out of the product we sold, then use all that poundage to add a delivery charge. I reasoned that not only would that enable me to move seamlessly between subcontractor and employee, but it would also allow me to monitor the delivery enterprise and make sure it financed itself.

I'm convinced that not a single farmer who does delivery and imbeds the cost of delivery in the product is covering all the delivery expenses. The farm inevitably subsidizes the delivery business. As farmers, we desperately want the sale and we're willing to drive to town, put it in the oven, on the spoon, and feed it to the customer if that will create a sale. All those amenities cost money. They take time, fuel, and equipment.

In *YOU CAN FARM* I have the rate sheets for this, but it was brand new at the time and a fairly unproven idea. I can assure you now, after many years, that this is one of the most brilliant ideas I ever had. It has indeed enabled us to move seamlessly from subcontractor to employee and back again--a few times. It enables us to track our delivery service as an independent business. It's the basis of our commission for Richard if he exceeds a volume benchmark.

Always, always, always separate your delivery charges from your product charges. It may seem a little cumbersome, especially to the customer, and especially to chefs, but it creates a clear picture for both parties to understand what the true cost of distribution is. Both parties being more knowledgeable is a good thing. Never apologize for it; always explain it in terms of more comprehensive information and transparency.

This delivery separation has two problems. The first is that it is price prejudicial toward low-priced goods. A 25-cent per pound delivery charge on filet mignon at $25 a pound is a spit in the ocean. But a 25-cent delivery charge on basic lower-priced Kennebec potatoes at $1 a pound is a huge percentage. What's the answer?

If you go down to the UPS store and send two boxes, each weighing the same amount, do they give you a discount if one box is feathers and the other diamonds? Of course not. They go for the same rate if they weigh the same. The answer to the dilemma? Simple: if UPS can't figure it out, neither can I.

The second problem with this delivery scheme is that it can

create misperceptions when a client steps up their purchases. We do have a sliding scale based on pounds, but it never goes to zero. Some delivery services charge a flat rate per stop and don't weigh anything. That's a defacto sliding scale too, inasmuch as it's prejudicial against smaller clients. Let's say we start with a chef who gets a case of eggs a week. She pays $14 for delivery.

A couple of months later, she begins ordering two cases. Once she started using our eggs, of course, her customer satisfaction went up, everyone started coming in, and local buzz went through the neighborhood because she was using Polyface eggs. Come on, enjoy the story with me. So we take her two cases of eggs and the delivery is $26 (not $28, which would have been double the $14 single-case rate), a drop in the rate but a doubling of the poundage. She calls upset: "You've been bringing me one case for $14 and now you charge me $26. It doesn't take any more gas or time to drop off two cases as one. Why the big increase?"

You know what the answer is? Okay, let's say you take two packages down to the UPS store, and both are addressed to the same place. Does the lady behind the counter say: "Oh, I see both of these are going to the same place. Well, since we're going anyway for the first package, we'll take the second one free." Is that what she says? No way. So the answer is: if UPS can't figure it out, neither can I.

The take away for this is to appreciate that in marketing and business things happen that on the face of it seem unfair. Don't deny it or downplay it. Step up, admit it, recognize it. Deal with it head on. Your customers will appreciate your candor and openness. And sometimes copying is the best you can do.

Always, always, always separate your delivery charges from the product charges. One final thing this does: it allows you to simplify your pricing. If you want your on-farm pricing to be the cheapest, you create that simply by adding the delivery to those customers who want it delivered. They're both working from the same price sheet, which is simpler for you and actually more informative for them. If a very large customer you deliver to wants to come and buy bulk and save that $50 delivery fee, she knows exactly what the prices are.

The separation is key to running a delivery enterprise that stands alone.

5. Everyone food. When creating a product mix, remember that if you're going to stay in business, you have to offer things that people use. Exotics are fine, but it's hard to build a thriving sustainable business on just exotics. For sure, what might be exotic in one place may not be in another. The average American eats only .7 pound of lamb a year, so in most places lamb is fairly exotic. But in some ethnic areas, it sells better than beef.

Just realize as you put together a marketing plan that exotics have limitations. This is what ultimately killed the ostrich and emu fad many years ago. A few enthusiastic folks got into it and infused others with enthusiasm but as soon as the most venturesome consumers had tried it, "been there, done that," the market crashed. While it was a fad, it looked like a good deal. But it flashed and burned pretty quickly.

Turnips are good, but most people don't eat that many. I wish I had a nickel for all the struggling produce growers whining about not being able to get people to buy, including chefs, but all they want to grow is heirloom stuff. At Polyface we supply 50 restaurants and the chefs constantly lament to us, "Where's the person who will just grow good compost-fertilized white potatoes and Better Boy tomatoes, Nantes half long carrots and California sweet bell peppers?"

They want to buy bags and bushels of these staples, but people who come to farming with stars in their eyes and doped up on specialty thoughts only want to grow $3 a pound heirloom tomatoes. Folks, a big market exists for $1.50 a pound, prolific, great tasting basic tomatoes. Use the heirlooms as garnishes. That's great. But what I call "everyone food" is steady, dependable, and stable for your farm business.

Ditto for heirloom turkeys. I'm glad people are raising them. But on our farm, we have a vision to eliminate factory farming. Period. It's a blight on the planet, unnecessary, and evil. I've got news for you; we're not going to eliminate factory farming with $150 dark-meat 12-pound heirloom turkeys. People only change so fast. Paul Harvey used to say "civilization at its most accelerated pace is agonizingly gradual."

As Allan Nation, founder of *The Stockman Grass Farmer* magazine used to always say with his characteristic charming humor, "Your customers will only let you be so weird. You can be a nudist, or you can be a Buddhist. But a nudist Buddhist is just too weird." And of course he was right. People change slowly and they're afraid of unfamiliar things. Goodness, I'd love to see the day where the double-breasted Cornish Cross broiler became a distant memory. But it's our cultural context, and before we can get people weaned off them, we must first get the chickens out of the factory farms. One step at a time.

Marketing food that most people already eat is far simpler than trying to get people to buy something with which they are completely unfamiliar. Offering a better iteration of a product they're used to is a much easier sell than a completely different product.

6. Find your fit. All marketing options will not mesh with your goals or your vision. Realize that you will have to pick and choose among opportunities to find what works financially, emotionally, and logistically for you and your operation.

For example, here at Polyface we created a set of criteria to protect us from markets we would later regret. The two key ones were that we would not sell to someone who put their label on our product and the second was that we didn't want to sell to an outfit who would ship our product out of our bioregion.

The one criterion preserves our brand integrity and the second preserves our local-centric identity. We have some others related to these, like not delivering product more than four hours away from the farm. Of course, if you want to drive from a thousand miles away and get it here at the farm, that's great. But we won't ship it to you.

We're beginning to struggle with that now that we have shelf stable jerky and this may change even by the time this book actually comes out. These kinds of things are fluid, but it's healthy to wrestle with what you want your farm business and its marketing persona to look like. Some people want to stay extremely small. Others are happy to grow an empire. Marketing fit is quite different in those two scenarios, and it's best to deal with them earlier than later. Here at Polyface, sometimes a market opens up and we jump on it, only to find out a year or two later that it's not a good fit. If we had wrestled with the ramifications of that market before jumping in, we would have saved ourselves some heartache and the extrication process later on. Marriage is a lot easier than divorce. You can get pulled into what may seem like the latest greatest opportunity only to rue the day later on.

Don't just assume that because somebody wants your product it's a good idea to sell to them. Run it through a matrix of heart-level objectives and make sure it's consistent with your true north.

7. Cash flow. I've made the point already that most businesses do not fail due to lack of service or product; they fail due to cash flow. This is doubly acute in fledgling or expanding businesses; they're always starved for cash.

I'm writing this book on a MacBook Pro and I remember the first computer we bought, back when all the forecasters were predicting the end of Apple. You young people reading this today should look back at those archives from the 90s when dirges for Apple were everywhere.

I remember well considering PC and Mac and making the decision for Mac. One of the big reasons was that I didn't want to lose choice in the marketplace. I didn't and still don't know a thing about computers. I can barely turn it on. I write and do email, but that's about the extent of my computer savvy, and even then, if something strange comes up, I melt down. I mean melt down. I'm paralyzed.

Anyway, I'm happy today that I bought my first Mac when the company was on the ropes. I'm sure that purchase saved the company in those financially scrapped days. Underdogs, unite! That's my cry. But look at Apple since. It's flush with cash. And

Direct Marketing – How

it might need that cash to ride out continued business cycles. I'm hoping that some Apple executive reads this paragraph and decides to send me a $1,000,000 check for saving them when they were in the doldrums.

One of the most important things to think about in marketing is cash flow. What can you sell every week? What do people buy every week? What can you produce every week? How can you minimize down time or "we're out of it" time? These are all part of your marketing strategy. What products have a higher turnover?

On the production side, this is certainly one reason why produce farms are common. From seed to harvest is just a few weeks. Compare that to Iberico ham that takes two years to grow on the pig and then 3-5 years to cure in special caves. No wonder it costs $100 a pound. That's a long turn-around time in production. Once the pipeline is full, great. But getting there can be problematic.

That's one reason why I like broiler chickens. It fulfills all these criteria. America is eating chicken (per capita consumption is now higher than pork and beef), so it's an everyone food. It's a consumable. And it has high turnover. Start to finish is only eight weeks. That's almost as fast as a radish.

Maintaining market continuity is critical to build loyalty and trust. If you're in and out of inventory, it's hard to build a reputation. Of course, if your whole narrative is seasonal celebration and you grow strawberries, that's fine. The time you would otherwise spend dealing with customers year round you'll invest instead on publicity and building hysteria coming into the new season. And if you're really smart, you'll figure out how to make strawberry preserves and strawberry pies from frozen berries that you can sell in the winter time. Cash flow is a big deal; write it into your marketing and production plans.

8. Gateway products. How does a customer get in? As expendable cash becomes tighter and the personal savings habit dwindles, offering customers a cheaper way in is more important than ever.

Anyone selling sides of beef or pigs will find that the day you offer pieces, sales escalate dramatically. What's the cheapest thing a customer can buy to try your farm products? We have to think incrementally.

This is one reason I like eggs. I call eggs a gateway product because they are an easy way in. Everybody uses them; they come in small quantities; they are a complete package. From an inventory perspective, nobody buys a partial egg. The problem with piecing out, of course, is that now you have inventory problems. You trade sales volume for inventory juggling.

Every pig is not ribs and every chicken is not breast. But it's amazing how many people will pay $14 for a breast who balk at paying $15 for the whole chicken. Why? Convenience. Ease. The whole chicken is more than the family can eat in a meal, so you have leftovers, or waste, or both. The breast is a portion that can be parceled out to fit the number of diners.

Half of all Americans right now cannot put their hands on $400 in cash. That means a $500 CSA share or a quarter of beef is not marketable to half of the country! This is why we have to break down what we sell into gateway items. What takes the least fussing over? What is cheapest?

Sometimes the gateway can be an expensive item that's microwavable. Don't chastise me for saying that, you purists. Would you rather have a pasture-based GMO-free compost-grown pot pie from a microwave or a factory-farmed chemical-based antibiotic-laden pot pie from a microwave? Give me a break. If we're going to put it in the microwave, we may as well start with something better. We crawl before we walk, and many of our customers literally have to learn how to crawl.

What's your gateway product? Identify it and then promote it front and center. Show people how good it is, how to use it, how to buy a bunch of it. Once they're happy with that one, they'll come back for the other things.

There you have it, my cornerstones for an effective direct marketing farm strategy. Some day I hope to write a book titled *RELATIONSHIP MARKETING* in which I'll drill down deeper into all these concepts, but this is definitely enough to start. As I've put these down, I've realized I need to do better at my own advice. Isn't that always the way it is? We know stuff, but we don't do what we know. Oh well, let's enjoy this marketing journey together. It's by far and away one of the most exciting aspects of the whole successful farm venture. To grow it is one thing. To build relationships with folks who appreciate it and love it, and in turn shower gratitude on us; that takes our farm to a new level. I know how important it's been for us, and I wish you that same level of blessedness.

Chapter 9

Gross Margin Analysis

This chapter is about money. Knowing where it is, comings and goings, how much was won or lost. Although farming is more than a business, it is still a business and needs to be treated as such. Financial hemorrhages devastate small and fledgling businesses.

I'm desperate to keep this chapter from being tedious because often money matters and accounting tend to be that way. I'm hoping I can present it in an exciting way with real applications to real situations. I do a lot of conference speaking, and it's always interesting how the two topics farmers shy away from are marketing and money. Every time I do a session on either of those, inevitably the question period turns back to how-to: how to build compost, move the cows, build fence, construct a pond. For some reason, we farmers avoid marketing and money like the plague, and yet they are foundational to our successful farm businesses.

If we don't make a profit, we don't stay in business. If we don't know our financial position, we won't make good decisions. Tracking the money is as important as anything else on our to-do list.

It all starts with bookkeeping. My dad, who was an accountant, referred often to a friend in Venezuela who, in broken

English, would admonish, "Senor Salatin, you 'ave to pud id down." Dad always used that admonition when clients balked at meticulous and categorized record keeping as too tedious. The point was that if you don't write it down, in a clear record, in a designated place, you can never retrieve the information.

So it all starts with a financial record keeping system that makes sense but mainly contains enough specificity to be able to tease out lots of different threads. Most mainline regular record keeping systems are not specific enough. You need to customize it to your situation. I've already mentioned that we use Quickbooks. Back before computers, we simply used a ledger. Then we graduated to Quicken, which we found similar and intuitive. But when we hired a bookkeeper, she wanted us to convert to Quickbooks because that's what she was familiar with and it's configured to be tax-preparation friendly.

The point of all of this is I don't want to get bogged down on the system. Plenty of options are out there. What is necessary is to configure the system so it provides meaningful and clear information. The whole reason for accounting is to be able to track the flow of money. Let's give an example. Let's say you have a produce farm. How are you going to know if you're making money on carrots?

Assume you're growing carrots. If I asked you, "Are carrots profitable?" would you know how to answer me? In the next chapter we'll get into time and motion studies, which is related to money but far more. That's where we add stopwatches to the accounting. Right now, hang with me on the money.

How much does the seed cost? How much do the carrots bring at sale? If I just have an "Income" category and lump all my sales together, how do I know how much the carrots I grew were worth? Unless I keep carrot sales separate from cabbage sales separate from cucumber sales, I have no idea which segments are profitable and which aren't.

On our farm, we raise several different kinds of poultry: laying hens, breeder hens, broilers, heritage roosters, pullets, and turkeys. If we simply had an accounting category labeled "Poultry

Feed" we could lose money hand over fist on the laying hens but not know it. The secret to a functional accounting system is categories. It may seem anal, but if you have information, you can make good decisions. If you don't have information, you'll tear your hair out in frustration trying to figure out where money is lost.

When I was a teenager we had a friend who started a small abattoir. He lived a couple of hours away so had a local accountant set up the books for the business. A couple of years in, the business was going broke but nobody could figure out why. Dad had roughed out some categories for the friend in one of their discussions before launching the business, but the local accountant scoffed at Dad's tedious categories. He discarded many and lumped them together in order to make record keeping more efficient.

In desperation because the business was going broke, the friend called Dad in to help. Immediately Dad realized that the simplified record keeping of the other accountant did not offer enough information to find the hemorrhage, so he redid the books. Within three months they found the problem but it was too late to save the business. As it turned out, the delivery driver was running a side meat business, stealing a little from every load, but the categories did not separate delivered from in-house stock. As soon as they had more categories and could do some internal reconciling, they found the problem.

Most businesses have multiple income streams, each of which has an associated expense matrix. If we don't separate expenses between laying hens and broiler chickens, for example, we'll never know the financial health of either. Even though each one may have identical type expenses like feed, chicks, propane, equipment, and labor, all of these categories must be repeated for the other enterprise. Here at Polyface, we have broiler feed, broiler chicks, broiler propane, broiler labor. And then we have laying hen feed, laying hen chicks (pullets), laying hen labor.

To be sure, you can't segregate every single expense. Those are called overheads: property taxes, insurance, vehicle, tractor, rent, mortgage, administrative salary. All the economic sectors have to pay overheads out of what's left over from gross margins. Your direct costs are expenses associated with particular enterprises; explained another way, they are costs that would go away if you quit that enterprise.

For example, if you quit raising laying hens, all costs associated with laying hens would go away. When figuring enterprise gross margins, these are the figures we're dealing with. We're not putting in fixed costs, or overheads, because they would occur whether we were doing that particular enterprise or not. If you quit raising laying hens, you would still have to pay farm insurance, property taxes, and perhaps own a farm truck.

To review: Enterprise income minus direct cost equals enterprise gross margin. On our farm, we have 200 categories and generate nearly a 10-page profit and loss statement. While that may sound unnecessarily tedious, when I get to the end of the year and want to look at performance on each enterprise, I can easily tease out the information and get a margin.

Remember, the cumulative total of all the margins has to pay the overheads. So even though a particular sector may generate a profitable margin, that doesn't mean the business is profitable. It could be that other enterprises are running at a margin loss, or it could be that the overheads are too much. You only have a real profit when all the direct costs and all the overheads, or indirect costs, are accounted for. This is why overheads like your dwelling, your salary, your automobile, the tractor, taxes, and insurance are so important.

It doesn't matter if your egg laying operation generates a $2,000 gross margin every year; if your salary is $20,000 and the truck depreciation is $2,000, you either need to work for nothing or add some other enterprises or increase the current egg operation. On most small and medium-sized farms, multiple enterprises are necessary to cover those overheads. We'll talk much more about that in a couple of chapters. For now, realize that business fairies and witches do not exist. It's up to you to figure out where you are and figure out your weak links. And the main point here is that you'll never figure it out until you have enough figures in the right categories to create meaningful equations. If you don't have it, you can't fabricate it or remember it. You have to put it down.

Now let's see how this works in a real practical scenario. Let's say we're running a grass-finished beef operation and we're trying to figure out whether to buy calves or keep brood cows. It's a complicated issue because if we don't keep brood cows, we can sell a lot more calves since we don't have to keep depreciating stock. But if we keep brood cows to birth our own calves, we don't have to buy calves. This is a question with which every person who's ever direct marketed grass finished beef has to wrestle.

In order to figure that out, we have to know the cost of keeping the cow compared to buying a calf. Ultimately, before any profit can be made, the calf has to pay for the cow's upkeep. She's a depreciating asset and she eats a lot. Is it better to let someone else own that cow and produce that calf, or is it better for us to do it? In order to figure out the cost of keeping the cow, we need to decide what our grass is worth per cow-day. A cow-day is what one cow will eat in a day. All the meals you're going to eat today amount to one person-day of food.

By converting to a constant, like inches, bushels, or cow-days, you have a standard measure for comparison purposes. So we take all of our cow-days in a year and create a total. That means converting calves, stockers, bulls, and finishers to cow equivalents and multiplying it by days. Most cattle operations have a variable number throughout the year. We might have a bunch of finishers in the spring that get processed in May and June, which drops the cow equivalents. If we're calving in the spring, obviously adding a bunch of calves increases the cow equivalents (7 calves per cow for the first month, 5 calves per cow the second month, 4 calves per cow the third month, and so on).

Let's say we have 50 cow equivalents from January 1 through April 1. That's 120 days X 50 cow equivalents is 6,000. Write that down. Then perhaps we have 70 cow equivalents April 1 through June 1. That's 60 days X 70 cow equivalents is 4,200. Write that down. Then maybe we sell some heavy finishers and drop back to

Gross Margin Analysis

50 cow equivalents from June 1 through August 1. That's 60 days X 50 cow equivalents is 3,000 cow-days. Write that down. By now the calves have grown quite a bit so now we're up to 70 cow equivalents for August 1 through October 1, or 60 days X 70 cow equivalents, is 4,200 cow-days. Write that down. Now we cull a couple of cows and sell the bull but the calves are much larger, so we stay even at 70 cow equivalents October 1-Dec. 31, or 90 days X 70 cow equivalents is 6,300 cow-days. Write that down.

Now if you add up all the cow-days in the year, it looks like this:

6,000
4,200
3,000
4,200
6,300

That gives us a total of 23,700 cow-days for the year. That's the total feed production off the pasture. Obviously if some of this is fed as hay, it just rolls right into the total numbers. You don't have to count bales; you have to count actual cow-days because that's what you're actually feeding. Cows will eat widely different amounts per day depending on whether they're lactating or not, whether the forage is drought-stressed or washy, and whether the weather is cold or hot. They eat more when it's cold than when it's hot.

The beauty of cow-days is that it automatically figures in these variables and creates an accurate picture of what you actually fed, or pounds of carrying capacity. Now that we have our total cow-days (remember, this is just like bushels to a grain farmer or board-feet to a forester) we need to find out what each one is worth. The income is easy. What did you sell? If your beef income is $25,000 for the year, that means each of these cow-days is worth $1.05 ($25,000 income divided by 23,700 cow-days).

But that's not our margin; that's only the gross value of each cow-day. To get to the margin, we have to figure all of our expenses. This is why we have to categorize. Obviously we have to make some assumptions, but the more we have down in black and white the more accurate our figures will be. So let's look at the expenses. In our purely hypothetical example, let's say our mineral is $1,000 (not mineral for chickens, or goats, or pigs, but mineral for cows--see why we have to keep it separate?). Hay--we fed for 60 days, 3,000 little square bales, and those cost $1.50 apiece (add up your fuel, a cost-per-hour on the machinery, depreciation and to be fair you need to make these up if you did it all yourself).

If you want to be imminently honest, force yourself to use the going commercial rate in your area. If you had to go out and buy that hay, what would it cost? Obviously you're going to price similar hay to what you made; cows don't need horse hay. Okay, 3,000 bales at $1.50 is $4,500. Now here we'll assume that the cost of acclimating calves only (bringing them in, getting them settled, electric fence trained, etc.) is equivalent to calving your cows. That's a labor wash, so we'll just let that ride.

What else do we have? Land cost--Dave Pratt at Ranching for Profit schools admonishes us to put in the going rent in your area whether you own the land or not. This keeps it fair. Otherwise you're living in la-la land; that land has a cost and it needs to be figured in even if you inherited it or purchased it with off-farm income. At $40 per acre rent, let's say the rent is $3,000. Now labor. This now dovetails with time and motion: how long does maintaining this herd take every day? For sake of discussion, let's assume one hour per day, which is 365 hours in a year, worth $15--is that okay?

Again, my wise dad used to say to his farmer clients who routinely lost money: "You might as well do nothing for nothing as something for nothing." If you don't make any money, why continue? Or why continue doing that; why not go and do something else that does earn you some money? So 365 days X 1 hour per day at $15 per hour is $5,475. If you're staying with me, we now have three major expense items written down:

Mineral	1,000
Hay	4,500
Land	3,000
Labor	5,475
Total	**13,975**

To be sure, most outfits would have a fertilizer component, perhaps some seed, veterinarian services, hauling, and some depreciation or at least maintenance costs on buildings (shed to store the hay in, to store the baler in, whatever). Chances are that among all these elements, we're going to have another $5,000 for a total direct cost of $18,975. Once we subtract that from our $25,000 gross income, we have $6,025, or .25 per cow-day ($6,025 divided by 23,700 cow-days). That means every cow-day gives us a gross margin of 25 cents.

Looked at another way, every cow-day costs us 80 cents ($18,975 total expenses divided by 23,700 total cow-days). One cow is 365 cow-days, so if we multiply 365 by 80 cents, we have the cost of keeping a cow: $292. But wait. If we're buying weaned calves, we also have the cost of that calf up until weaning, which is going to be roughly a third of a cow equivalent for 240 days assuming an 8-month weaning date. Multiplying the nursing calf for 240 days X the 80 cent per cow-day expense rate is $192. Add that to the cow cost of $292 and you come up with a total value of a weaned calf: $484. If we can buy similar calves for under, say, $450, it makes sense to do that and keep few cows or no cows at all. At least now we have a benchmark so we can make a decision.

Now we can go back to our cow-day composite and convert all those cow-days to growing stock to create a different cow-day margin scenario and see what things would look like without cows.

Before we leave this discussion, let's leverage this cow-day margin to one more level. If our margin is 25 cents, then that means anything we do to increase cow-days must carry that margin. If it's worse, we may as well do more of what we're doing. If it's higher, we might want to invest in it.

You see, dear folks, the problem with decision making is that all the people selling things do it compared to nothing. Don't get me wrong; I'm glad people are out there making things, inventing things, fabricating things, producing things. But our problem is in deciding what to buy among all the alternatives. The gross margin analysis gives us a better framework on which to make that decision.

175

Let's imagine that we go to a sustainable agriculture conference and trade show with our significant other. We visit the trade show and dutifully collect 30 brochures from as many different vendors who all have the perfect product for us. We watch the looping videos at each booth, handle the product, smell it, eat the Tootsie Rolls in the bowls, and listen to the sales pitch. Each one has charts showing the benefits of the product. What's interesting is that it's always compared to nothing.

It's not compared to the silver bullet offered in the adjoining booth, that vendor over there. Or the one across the walkway. It's always compared to nothing. Well goodness, almost anything will respond if you offer something versus nothing. We put the vendor's material in our little cloth bag stamped with the conference logo, and move on to the next vendor. We work our way around the trade show hall and finally exit, head bursting and tote bag bulging. Time for a break.

We go up to our room on the fourth floor and spread all the newly-acquired materials out on the bed. The problem with all this is that you and I don't make decisions based on something versus nothing. It's always something versus something. And it's usually not very clear what the right something is.

Wouldn't it have been wonderful, when you were courting your significant other and finally made the decision to say "I do," if all the alternatives had been terrible? Wouldn't it be wonderful if these things were cut and dried? Only one desirable mate in the lot; the rest are all ridiculous. It's not that way, is it?

How about when you were choosing a college? Wouldn't it be neat if all of them were terrible except one? Then the decision would be easy. Or how about your current job or vocation? Wouldn't it be neat if you were not interested in anything on the planet and could not imagine doing anything except one thing, and that was the only thing that offered itself to you? But life isn't like that.

You see, we seldom if ever compare something to nothing. It's always something to something. So as we work through the stack of brochures from the sustainable agriculture trade show and recollect the conversations, more times than not we stuff everything back in the tote and decide based on "I really liked that guy from XYZ. He just seemed open and honest and I think we'd like working with him." Done. Decided. Sealed and delivered. Cha-ching, sale made. But is this the right way to make the decisions that will determine our future farming operation? Of course not. It's laughable, but the truth is we have to laugh at ourselves because we all know this is how most of us actually make our decisions. I know I'm guilty.

So by what criteria do we make the decision? After all, we're not the federal government with limitless finances. We actually have to pay our bills and try to act like honest people. We can't patronize everyone; we have to pick and choose. Armed with our margin analysis, we can try one or two options and test them against our 25 cent figure. If a foliar fertilizer, for example, costs $40 an acre to apply, the additional cow-days had better pass the margin test. If we're averaging 150 cow-days per acre at a margin of 25 cents, that's a $37.50 break even benchmark on any investment. If that $40 an acre material doesn't double our production, it's not worth it. If it only increases our production by a third, the margin on those additional cow-days is worse than our current situation. If we can double production by renting more land for $40 an acre, we may as well do that.

For sure, some things won't give immediate results, but few things do. This is why I've always thought it would be a good thing for everyone who sells organic fertility boosters or systems to each put $10,000 in a pot, creating a war chest of perhaps $300,000 that could finance a true product-to-product comparison. Get 30 acres and apply the material or technique on one acre per vendor (assuming 30). Hire one crack researcher to measure some benchmarks.

We could measure soil carbon, biomass production, earthworm population, spider population, and pollinator population. A matrix of five things should be enough to get a rough idea. Run the test for 5 years and see what happens. Ideally it should be done in several different parts of the country because what might work extremely well, or be the weak link in one area, might not be in another area.

In his arid Australia, P.A. Yeomans said the weak link was always water. Well, you can test that on your own farm. We did. Many years ago we got some little cheap garden sprinklers from Lowe's and some pipe and a simple $300 pump, stuck it in a pond, and irrigated a couple of acres. By measuring the additional biomass and doing a cost analysis on excavation in our rainfall, we figured out that on the average year we could double production for a total capitalization cost of $750 per acre. In other words, we could excavate ponds to hold enough water to double production at $750 per acre. While that sounds like a lot, realize that land around here sells for nearly $7,000 per acre.

So if we could come in with an excavator and build ponds for $75,000 that would double the production on 100 acres, that's equivalent to buying land for $750. That's a lot cheaper than $7,000 per acre, or $700,000. How many farmers expand in acreage at a big cost when they could simply do better at hydrology and get more bang for the same buck? And remember, that's a one time cost. It's an expenditure that's good for probably 100 years. Prorate that out to a per annum cost, and it's $7.50 per acre. Now suddenly that foliar fertilizer or that magic bullet from Vendor A at the trade show has to compete with double production at $7.50 per acre.

I'm well aware that we have labor to account for in this comparison, and that will probably run it to $15 per acre, but many soil amendments cost way more than that. In grazing programs, this idea is similar when applied to moving animals. In our cow-day scenario, we had a 1 hour a day labor component. Some people reading this might wonder why in the world somebody would spend an hour a day on this small herd of cattle. I'm assuming we're moving them once a day, so we're setting up front fences and back fences each day.

I feel your push back. Let's assume we move them once a week. Are we going to check on them every day? A good stockman would. Just checking them is going to take half an hour because the cost is in being there, so the savings is not as big as it sounds. If we save 3 hours a week in labor spread over the non-hay feeding time (305 days) that's 130 hours per year in savings, at $15 is almost

$2,000 per year in labor savings.

But if we shift to once a week moving, that'll cost us at least 30 percent in productivity due to inefficient grazing. That means our 23,700 cow-days just dropped to 16,590. That makes an interesting equation when we compare our above scenario of 23,700 cow-days at a cost of $18,975 to this weekly move scenario of 16,590 cow-days at $16,975 ($2,000 less labor expense). Our 80 cent cow-day cost just jumped to 98 cents. Suddenly that weekly move doesn't look so good, does it?

Staying with our original cattle operation, what if we added an hour a day for 100 days to double the cow-days with irrigation? Suddenly we jump to 47,400 (23,700 X 2) cow-days by adding another $1,500 in labor. Even if we add another $3,000 in pumping, electricity and depreciation with a New Zealand K-line irrigation set-up, that only adds a total of $4,500 to our total expense bottom line. So now we have an additional $4,500 expense on our $18,975 for a total expense of $23,475 to generate 47,400 cow-days, which is a cost per cow-day of only 50 cents ($23,475 divided by 47,400 cow-days). How about them apples? We dropped from 80 cents per cow-day to 50 cents per cow-day by adding 100 person-hours a year with strategic water application that cost less than $10 an acre to install. By investing in irrigation, we changed our margin per cow-day from 25 cents originally to 55 cents. Pretty big change, huh?

You see, dear folks, I'm convinced that Darren Dougherty of ReGrarians is right: the hardest climate to change is the climate of the mind. While we're all worrying about climate change out there somewhere, we need to be much more concerned about the climate of the mind. The only way we can change the actual climate out there is to first change the climate of our mind.

Now I hope some of the principles are adding up. One of the reasons to develop eclectic awareness is so that when we run marginal analyses we'll have a full toolbox of possibilities. We're constantly making choices between A, B, C, D, and E. But what if the answer is really F and we've never heard of F, or never even thought about F? What if F isn't even in our paradigm? This is why we must come out of the box, out of the shell, to explore and inquire of people who are like us, people who aren't like us, and everything in between.

This whole analysis speaks to opportunity costs. That's the value we give up by investing in something. In other words, since we only have a certain amount of money, what am I going to miss by placing it here rather than there? If we stay with our little cattle herd for one more illustration, imagine asking the question: should I excavate a pond for irrigation or invest that in a food truck? If irrigation will cost me $50,000 to double production but that same money invested in a food truck enterprise will enable me to move my income from $25,000 to $50,000 due to value adding, what is the right decision?

Just for fun, let's say that this investment would get our little cattle operation the cow-day increase in the example. If you'll check back, you'll see that the irrigation dropped our per-cow day expense to 50 cents from the previous non-irrigation 80 cents. So now we'll keep our original expenses ($18,975) but move our income up from $25,000 to $50,000. That gives us a total value of not $1.05 per cow-day, but a value of $2.11 per cow-day. Now we subtract the original 80 cent per cow-day cost and that leaves us with a gross margin of $1.31, which makes the irrigation at 55 cents (remember when we were ready to call the excavator and go for it a couple of paragraphs ago when it appeared worlds better than 25 cents?) look like the stupidest thing in the world.

Yes, I realize I'm running a lot of abbreviated ideas here. Of course someone has to operate the food truck--but somebody had to operate the irrigation. Look, if you want to argue, we can argue. But don't miss the forest for the trees. We can quibble all day; the point of this whole exercise is to follow an idea through different pathways and go through the protocols of margin analysis to good decisions. I cannot emphasize enough that what we've just gone through is the nuts and bolts of success and failure for your farm venture. It's that important.

What will give the greatest marginal reaction: this or that?

We have a certain amount of time and a certain amount of money. You can't just intuit this stuff. I don't know how many times I've known how an analysis would turn out, only to be completely surprised at the bottom line. Things you wouldn't think would have much affect at all can swing a business wildly toward profit or loss. And things you'd think are really important might not make a hill of beans difference.

Going through the process is how you create the confidence in your decision. This is called Working On The Business (WOTB). As Ranching For Profit schools emphasize, most of your outcome is determined by WOTB, not WITB (Working In The Business). WITB is what we spend most of our time doing, but it's actually the least critical. As Stan Parsons used to say, "we've become incredibly adept at hitting the bull's eye of the wrong target." Dave Pratt takes that one step further to explain the difference between efficiency and effectiveness.

They are not the same. We can be efficient at the wrong things. Just because you're efficient at planting beans doesn't mean you're effective. Perhaps you shouldn't be planting beans. Discovering why is being effective; that's WOTB. And that's where you go to your desk, calculator in hand, financial records open, and start pushing the pencil.

Let's finish up this margin discussion by talking a bit about pricing. We spent most of our time talking about playing around with expenses and very little with the income edge of the margin equation. But obviously you can increase a margin by either raising the price or reducing expenses. How do you set prices? Obviously, you have to know what your margin is in order to set a price. That circles back to good record keeping.

Beyond that, the science of price pointing is quite interesting. The rule of thumb is that if 20 percent of your customers are not complaining about price, you can raise it. Most of us can't handle 20 percent negativity and never approach this threshold. But half that rate, 10 percent, is certainly doable. Most of us never hear a complaint, or at most only seldom. In fact, most of us are embarrassed about prices. If you routinely hear, "Oh, that's one of the lowest prices I've ever seen," then you can be sure, raising prices is doable.

I'm not a pure capitalist. I suppose a pure capitalist would simply say your price should be whatever the market will bear. But if that's the only criterion, it will skew other considerations, like market penetration. For example, I know a guy who ships eggs into New York City overnight FedEx for $15 a dozen. I have no problem with that and obviously the market can bear it.

But my life's vision is to first serve my neighbors--that's not all I'm about, but it's a healthy portion. I can't think of a single neighbor who would pay, let alone could afford, $15 a dozen for eggs. For me, the reasonable balance is to establish a labor cost--what am I willing to work for and be happy? That will change from one person to another. I see no value and no legacy in being the highest price around. That might just indicate you have the slickest marketing team. It may mean you have the most glitzy distribution or website. It might mean you're a modern day snake oil salesman.

I don't see anything inherently noble or desirable in being the highest price around. More than my income, more than my wealth, I'm first and foremost a member of the human community generally and the Augusta County community specifically. What do working class people around here make on average? That's my starting point. Then I take all the other expenses and build them out around the labor cost, which we'll talk more about in the next chapter.

That gives me a total expense per pound, per dozen, per bunch, whatever. With that break even, I have to add a profit margin and if I'm also marketing it, a sales margin. I like to look at equivalent products from competitors. I don't compare our grass-finished 80/20 ground beef with generic in-store brand 70/30 conventional factory farmed feedlot product. I compare it to organic or at least name brand product.

I always enjoy going to Whole Foods Grocery store to compare prices. I come out of there thinking I'm giving stuff away. A rough idea about the competition's pricing will let you know if you're in the ballpark. Obviously if I'm way under, I know I have

Gross Margin Analysis

some room to pad. If I'm over, I know I have to try to shave expenses or be ready to explain the difference.

Leave room for sales commissions and the cost of storing inventory and some spoilage. Don't be afraid to adjust a price to accommodate risk or personal reluctance. For example, when we finally decided to do chicken parts and pieces, our biggest concern was that we'd sell the breasts and end up with a freezer full of thighs, necks and legs. Because we didn't really want to cut them up, and we didn't want the inventory problem, we set a high price on the breasts.

In full transparency, we weighed a group of birds whole and figured the price. Then we cut them up, weighing each group: wings, thighs, legs, necks, backs, breasts. We timed how long it took us (we use $20 an hour as our benchmark labor rate) and doubled it . . . because we didn't really want to do it. Then we set the price so we could throw away everything but the breast and still get our money back. The result is that we enjoy doing it (I didn't say I was anticapitalist, just not a pure capitalist) and our customers are satisfied.

When we have slow sellers, we drop the price. If we have trouble keeping something in inventory, we raise the price. It's always a dance, this inventory issue, and you're never right in balance. You're either short or long, but you keep tweaking and watching and eventually you come to something that works. If we have a big snafu, we run a Buy One, Get One (BOGO) sale. Boy, that works.

Sometimes we run a sale for a couple of weeks only. The bottom line is that you have to move stuff. It's a lot better to move it at half price than throw it away. Obviously you can't stay in business doing that all the time, but an occasional sale to adjust inventory is better than being frustrated or paying for expensive storage. For sure it's cheaper than throwing it away.

The study of price points is the study of psychological thresholds in pricing, or break points. For example, the difference between 99 cents and $1 is enormous. But the difference between $1.19 and $1.29 is nothing. You can wiggle around quite a lot between $1.19 and $1.49 without any prejudice. But when you tip over $1.50, you've entered a new psychology.

The whole perception thing is big. I saw a guy at the farmers' market struggle to sell one pound bags of garlic for $5. After a couple of weeks, he came with quarter pound bags for $2 and couldn't keep them in. If you're quick at math, you realize he nearly doubled the price. But the perception was that since it was in a cheaper bundle, it was cheaper. This is gateway pricing. People are funny that way, so be sure to experiment and play around with your price points and your packaging, both type and size.

For the record, the single item we've found least price sensitive is the pastured turkey. Unlike chicken, turkey for Thanksgiving is a festive table centerpiece, a celebratory sculpture for the family gathering. Turkey bologna, however, is everyday lunch meat. Far more price sensitive.

A couple of years ago during our winter planning meetings, when we typically do our heaviest margin figuring, Sheri asked an interesting question: "What is our most costly sales venue and what is our least?" At the time, we had four basic sales platforms: on-farm sales building, metropolitan buying clubs, restaurants, and online gift shop.

First, we isolated all the sales in those respective venues. Then we teased out the cost of making those sales: phone time, email time, cash register time. With all these kinds of things, sometimes you have to make a best educated guess to fill in some gaps, but we had pretty good figures since we knew what our salaries and commissions were to maintain these venues. By dividing the total sales in each venue by the total cost to capture that sale, we had a cost per dollar of sale.

Three of the venues were virtually identical at 8 cents per dollar. One was way off at 13 cents per dollar: the on-farm sales building. As a result of that analysis, we devoted a lot of time and energy the next year trying to get more people out to the farm in order to make the cash register busier. When someone is working the cash register, they need to be busy; otherwise, it's just wasted salary. This is the make or break kind of analysis that drives decision making on the farm and points it in the direction of the weak link.

We're a small outfit and don't have all the sophisticated tracking and weighing devices of big businesses. But that doesn't mean we can go lackadaisically along and assume all is right in our world. Once a year we isolate a group of animals and track them from live weight to finished packages. You have to do enough to be statistically significant. Don't just do one animal. Do a random batch.

By weighing live, we know what our yield rate is from live to carcass. By weighing the final retail packages, we know our cutability percentage from carcass to take-home. Obviously some of the cuts may go wholesale to a restaurant account. That's okay. Include it at its normal value. The point is to have a normal, random, accurate depiction.

Once we have all these numbers, we can accurately price for different sales. Some people want a live hog for a pig barbecue. Some folks want a whole carcass. We know what the processing costs are for each of these stages, so we can tease them out and come up with equivalent pricing for the various degrees of pricing. We set a benchmark margin by subtracting all hog expense from hog income. If that doesn't give our benchmark margin, we have to adjust something. That margin is necessary to pay for indirect overheads. The annual audit of a composite group of pigs (we do the same with beef) keeps our pricing honest to what's actually going on in the field and the bank account. Absent that audit, we don't have a clue whether our profit margin is acceptable.

Just be sure to capture all your costs, be fair to yourself and your customers, and keep adjusting. I've been at this for half a century, and I'm still adjusting prices. I make mistakes, misread the market, and have to go back, pull out the accounts, see where we can shave, where we can pad. It's a never ending game, but that's part of the excitement and challenge. If it was a done deal, we'd get bored, wouldn't we? So don't be irritated that this book can't and won't give you all the answers. Take the principles and apply them; they work and you'll figure it out. If you take the time to figure, the figures will eventually make sense. Trust me.

Chapter 10

Multi-Enterprise

If you're reading this in its proper order, which is AFTER reading *YOU CAN FARM*, you'll see a chapter in that book titled Synergy, Stacking, and Complementary Enterprises. Thematically, this will follow suit but my thinking and application have progressed far beyond that basic permaculture concept.

I have to credit permaculture (and I hope they don't sue me for using the term, since I didn't put it in a copyright mark or ask permission) for truly codifying this idea. Of course, like all things permaculture (the thought and design process developed originally by Mollison and Holmgren) it's based on natural ecosystems. Nature does not use monocultures. Nature develops polycultures.

Highly functional ecosystems express complex habitats. Riparian, open land (prairie) and forestal habitats intersect in a wonderfully inter-relational assortment of edges. Increased flora and fauna occur where these great habitats intersect. Relatively few living things thrive in only one of these habitats. Most like two.

Think about a dragonfly hovering over a pond, hunting insects but resting on the weeds along the shore. The shore is the intersection of land and water. Think about a wild turkey, whose poults need a 28 percent protein diet. Mast and seeds that drop in a forest tend to be low in protein and high in starch. Where are these newborns going to get the special high protein diet they require? High protein exists in the insects like grasshoppers and crickets living in fields, not forests. And so in the wisdom of nature, wild turkeys don't lay their eggs until late enough in the season to insure that their offspring will have access to insects. These nests are along field edges. If you want wild turkeys, you need lots of intersecting field and forest.

The late Jerry Brunetti, who founded Agri-Dynamics, captivated many an audience explaining the relative nutrition found in fencerows and overgrown field edges. The shrubs and so-called weeds that proliferate in these unkempt edges, he showed, were generally far more nutritious than any crop. Indeed, many have medicinal qualities that help maintain herd health. Clean fencerows are not the mark of a good farmer. They indicate a dislike toward nature's economy, which thrives on diversity and a pattern resembling a mosaic.

Some people have even referred to proper ecological farming as mosaic farming due to the land patterns we create. Forest, open, and riparian areas create their own mosaic, to be sure. Imagine flying over what is today the U.S. in the year 1400. The first thing you notice is ponds. Lots of ponds. Up to 8 percent (nobody really knows for sure) of the land is covered in beaver ponds. How many beavers? Conservatively, 200 million. This beaver activity was not limited to northern areas; it extended into the arid southwest and the high dry plains of the north. Although those dry areas were never as lush as the northeast, they were far more lush and hydrated than they are today.

Then you see massive grazing patterns on the prairie. Allan Nation used to tell me that it would have looked like the whole country was fenced with electric fencing. Both lightning fires and Native American-lit fires burned massive swaths across the land. These fires not only controlled forest proliferation, but also freshened up the prairies when they became overgrown with carbon. Preferring these burned-over areas, the bison, which certainly numbered more than 100 million, grazed following these burn lines as perfectly as if there were fences. Of course, these bison, along with untold numbers of elk, antelope, and deer supported more than a million wolves, each of whom needed about 20 pounds of meat a day.

Incalculable numbers of birds resided on this pre-European landscape: passenger pigeons, prairie chickens, pheasants and waterfowl literally clogged the air. Old diaries talk about the birds blocking out the sun for three days and birds landing in the trees and breaking off all the branches.

As we stare down on this landscape, the beaver ponds, burning, and animal movement would present a tapestry of unimaginable variety--a mosaic. An acre of native prairie contained upwards of 40 different species of plants. The sheer expression of life in this patchwork actually produced more nutrition than the same land mass today, even with chemical fertilizers, hybrid corn, factory farms, and tractors. That should give anyone drunk on Monsanto and Tyson pause.

A successful farm mimics this multi-dimension, multi-enterprise idea. What's wonderful is that a profitable, productive, and pleasurable farm does not degenerate the landscape; it actually regenerates and enhances the landscape. Several years ago Polyface participated in a Smithsonian spider study to see what our mob stocking herbivorous solar conversion lignified carbon sequestration fertilization did to the spider population. In the insect world, spiders are the keystone species like the lions are in Africa.

Spiders are predators, so if we have a healthy insect population, we'll have lots of spiders. A weak spider population indicates ecological weakness overall. These researchers used a metal circle about a meter in diameter and a heavy duty vacuum to quickly suck up all the insects and spiders in the ring. Then they'd dig into the ground with pen knives to extricate the ones burrowing away quickly. It was quite an operation.

They also got permission to study nearby pastures on fields that were continuously grazed, which is the more orthodox management plan. These neighbors' fields were literally a few yards away from our boundary fence. Our numbers were off the charts. But not just spiders: the whole spectrum was a beautiful balanced matrix of diversity. In the adjoining fields across the fence, the researchers could only find ants and beetles; very few of the other four genus types in the study, and certainly very few spiders.

For the last couple of years we (Polyface) have participated with another Smithsonian-developed study called Working Landscapes. The idea is to see how proximate farms can co-exist, or even augment, healthy ecosystems. The five study areas are: soil organic matter, pollinators, birds, invasive species, and forage species. We're site 29, and the other sites being studied are places where farmers have allowed an area to revert back to wilderness. Here at Polyface, at least as of now (two years in) we're the only farm offering actual utilized areas for the study, not set-aside areas. It's been quite fascinating to see the ebb and flow during the different study times throughout the year and with the different land uses. For example, the findings are quite different if we have a fully expressed, seeded-out pasture just before grazing or mowing for hay versus that same field a week after mob grazing or mowing for hay.

But so far, our numbers look extremely good and we're happy to report that we have every single known type of bumblebee in the state. That's apparently unusual because the researchers made note of it. Our field edges provide abundant turkey nesting areas and we see far more now than we did 40 years ago. Ditto for deer, bear, rabbits, and squirrels. The point is that a mosaic landscape re-creates the diversified habitats of pre-European times when the ecosystem was arguably far healthier than it is today. Lest anyone want to debate that, just remember that soil was being created at that time faster than its erosion rate. That alone should be enough to get our attention.

Back in the year 1400, the abundance of what is today the U.S. was astounding. If you travel to Jamestown, the first surviving English village in America, you'll find yet more corroboration of this. Surviving diaries of the first Jamestown settlers tell about the Chesapeake Bay waters being clear enough to see down 100 feet. For as long as anyone alive today can remember, the water has been turbid and murky to the point where if you stick your arm in the water, you can't see your hand.

Historians were puzzled as to how this could be. They never had an explanation until they discovered the original fort. Teresa and I honeymooned to Williamsburg and Jamestown when we were married in 1980. I remember well the tour guide telling us that the original Jamestown fort was "out there." She pointed out into the James River estuary, off the point of land where the colony's buildings had been excavated. Archaeologists thought the river banks had eroded about 100 feet and the original fort was out under the water.

But lo and behold, in more recent years, they've discovered the original palisade walls. The wooden posts, of course, are long gone, but the wood-decayed cylinders are still there. Today, you can follow the wall. Most importantly, once they found the wall, they found the trash pile. That's where the treasures are--the trash pile. And in that pile, they found oysters nearly a foot in diameter. Oysters are one of the most significant purifiers of coastal waters--they're like kidneys in the sea. So these magnificent oysters purified the Chesapeake Bay and kept its waters clean. Or at least, that's the current theory. Did I mention the abundance in 1400? Can you imagine an oyster a foot in diameter?

Abundance requires many parts to work together, harmoniously, to produce a whole in which 1 + 1 = 3. That's abundance, synergy, symbiosis. Call it what you will, it represents regenerative systems. Good individual farming should actually increase the commons, not decrease them. As a result of our presence, hydrology should be better, not worse. Soil should increase in both volume and fertility, as opposed to erosion and depletion. Air should be more oxygenated, less carbon dioxided (not a word, but you know what I mean), and cleaner. Species should proliferate, not collapse. Biomass should increase, not decrease. This never happens in a linear, reductionist model; it only happens in a multi-enterprise pattern. That's nature's template.

Here's the point: a successful farm mirrors this pattern in all its dimensions. More enterprises make life more exciting. Instead of doing one thing year after year, or day after day, we get to do a variety of things. This diversity feeds the human spirit.

Multi-Enterprise

For example, because we not only raise chickens but dress them and sell them on our farm, in a given day, a person will move, feed, and water chickens in the field at morning chores, process chickens for a couple of hours between breakfast and lunch, bag chickens and interact with customers in the afternoon, and move a group of pastured pigs before supper.

Compare that to the day of a typical processing plant line-worker in an industrial plant. She stands all day, every day, in the same spot on the same line, no sunshine, no songbirds to hear. Chlorine-sprayed walls, concrete, and dead chicken after dead chicken. All day, every day. All day, every day. I find it reprehensible that the animal welfare certification groups who go to great lengths to make sure that people buy respected chickens do not even blink at the soul-sucking, dehumanization of workers in large-scale processing plants. The same could be said at tomato packing plants and most other food factories. These realities alone should drive the local food, diversified farmer model.

Because our farm embraces all these different enterprises and even multi-dimensions within enterprises, we have lots of activities to enjoy throughout the day, every day, day after day. Most discussions promoting multi-enterprise farms focus on the economics of stacking and layering income streams. Yes, I get that and we will get there soon enough. But I wanted to start by laying the foundation of what this diversification does to stimulate the psyche of a person.

Whether it's riding all day in the cab of a tractor, working in a processing plant, or picking cabbage all day in a California field, orthodox industrial farming fails first because it fails to feed the human spirit. If for no other reason than to feed our souls, successful farmers need a multi-enterprise approach.

As discussed in the marketing section, multi-enterprise farms leverage customers who enjoy the benefits of a one stop shop. The synergy with direct marketing is that adding enterprises is easy. A commodity-based farm has to find an entirely different market. If a corn farmer wants to grow green beans, the grain elevator won't buy green beans. The commodity farmer wanting to diversify has to find

191

an entirely different market, location, and network.

Imagine if a Tyson chicken farmer wants to start selling peaches. Tyson doesn't want peaches. An industrial commodity farmer can't leverage whoever is buying his stuff because the markets are themselves limited-commodity. Yes, a grain elevator may buy corn, soybeans, and wheat, but it's limited to grains. What if a farmer wants to grow something other than grain? Or a weird grain, like amaranth or spelt?

This is why direct marketing offers some leverages that might not be apparent at first glance. While the beauty of commodity markets is that you can generally grow as fast and as much as you like, you can't go too far off the reservation. The tomato processing plant doesn't want grapes. But with individual customers and smaller scale, a farmer can play around with lots of different enterprises without having to go through the hassle of lining up a new market. Most people who eat wheat are also interested in amaranth, chickens, peaches, grapes and tomatoes. The household purchase portfolio is far more diversified than a large-scale commodity buyer.

A multi-enterprise farm leverages not only the customer but also the land base. You can tuck things into micro-climates to take advantage of assets that would be impossible for limited-enterprise farms to access. For example, we have a hugelkultur bed on the south side (sunny side) of the hay shed, under the eaves so it catches the roof run-off. A PVC floating garden works great in the pond. Terraces along the hoop houses offer garden space.

A reflective, steel-sided structure coupled with a cold frame can offer magnificent season extension for growing plants, as do subterranean greenhouses and solariums on the south side of a house or outbuilding. The new science on tall tunnels is that the most efficient design is a vertical heavily insulated northern wall and then half a hoop on the south. The reason is because the north side (at least in the northern hemisphere) does not collect much solar gain, so it actually dissipates much of the gain collected by the southern half. By heavily insulating the north side to stop that heat transfer, you actually get a more functional space.

Multi-Enterprise

Right now, as I sit here thinking about this, I'll list all the enterprises at Polyface just to get a flavor of the opportunity and leverage I'm talking about. Be assured that this is far from what I think could or should be done.

We haven't even scratched the surface, but here's what's being produced right now for sale:
- Beef
- Pork
- Chicken
- Turkey
- Rabbit
- Lamb
- Chicken eggs
- Duck eggs
- Pullets
- Honey
- Elderberry juice
- Maple syrup
- Grasstains tours (school and corporate tours)
- Lunatic tours
- Polyface Intensive Discovery Seminars
- Lumber
- Firewood
- Vegetables
- Rustic wood products
- Burn-etched wood signs
- Rough cut artisanal log butt slabs
- Books
- T-shirts
- Post cards
- Seeds
- Handmade cosmetics
- Crocheted baby slippers

Just so you believe me when I say we haven't scratched the surface, let me share my list of possibilities for the future:
- Leather
- Fruit (apples, plums, mulberries)
- Bramble fruits (blackberries)
- Wreaths (grapevine, Christmas, etc.)
- Flowers
- Fish
- Freshwater crawdads
- Furniture
- Children's wooden toys
- Overnight accommodations
- Commercial kitchen (soups, stews, pot pie, stock)
- Dairy (milk, cheese, kefir, ice cream, etc.)
- Fur accessories (mittens, gloves, hats)
- Grapes
- Ferments (sauerkraut, kimchi, etc.)
- Mushrooms
- Quail
- Waterfowl
- Food truck

That's enough to keep me going for awhile, don't you think? If you think about it, the multi-enterprise farm is the foundation of building community. But what is the official word from the USDA? I like to say USDuh. The official word is that we don't want people on farms. The U.S. is wonderful because so few people have to stoop to the drudgery and peasanthood of farming. Just to make sure everyone understands this list, please realize that not only is our farm in a very different theme than the USDA's vision, it's equally opposite to the official environmentalists' agenda. Look at any conservation easement, any environmental trust document, and you will see an equally anti-human element. To be sure, they don't like industrial mono-species farms either, but so far these groups have not joined the multi-enterprise, integrated-economy of tomorrow's successful

Multi-Enterprise

farms. They're stuck in the 1950s paradigm and the notion that farms are really our nation's playgrounds.

But you can't have a functional food guild, a functional community, without multi-enterprises. Even if we agreed that the industrial agricultural community exists in the most abstract sense of that word, we can't feed ourselves without a highly productive multi-enterprised counterpart. You can't have people-centric multi-speciated farms and attract the best and brightest and be as productive as natural ecosystems without the multi-enterprise model.

If you're going to actually bring the bird component of the pre-European ecosystem back to its functional strength, you can't run cattle without chickens. It's okay to replace wild systems with domestic, but you have to preserve the loops, the synergies, if you're going to preserve similar ecological guilds and human communities. If we don't have the disturbance of the buffalo or fire in our forests and savannas, we need the pig to do that essential freshening up. The carefully managed porcine disturbance unleashes a latent seed bank to stimulate species that would lie dormant in an ecologically static state.

I call this ecological exercise. So if you're going to have chickens with the cows like birds with the herbivores, suddenly you need people and markets beyond just cows or just chickens. And if you're going to have pigs for proper disturbance, you need processing and marketing for those. If you're going to have produce, you need a fertility stream, a salvage stream, and a market. The more you simplify the farm, the less you follow nature's template. The more you diversify the farm, the more you follow nature's template but the more complex it becomes. Welcome to nature.

Vegetable production works best when combined with animals because manures are magic. One of the reasons there are no animal-less ecosystems is because animals spread around fertility. They also decompose biomass more perfectly and faster than compost piles. Their digestion is always the right temperature, the right moisture, and the right microbial mix. When people complain about animals being inefficient, of course they're inefficient. That's why they leave enough goodies behind to fertilize more than they extracted. That's why all the rich, deep soils on the planet developed under prairies with herbivores. They did not develop under forests or cultivation. Vegetables and grains, which are mostly annuals, deplete soil; they extract more fertility than they create. Perennials, on the other hand, feed soil.

Annuals store their excess energy in seeds and fruit (cucumbers, squash, pumpkins). They depend on this above ground energy savings account to survive and sprout another year. Perennials, on the other hand, store their excess energy underground. Whatever weather and devastation occurs above ground, the root reserves send up new shoots the next season. The difference explains why annuals are extractors and perennials are builders--the energy flow is opposite in the two types of plants. That is why combination systems work best; in other words, integrating annual-perennial, plant-animal, and other relationships yields not only more production, but more overall environmental stability.

Permaculture disciples use the word stacking to denote some of these ideas. On our farm, we put small pigs in a hoop house in the winter. With 4 ft. X 8 ft. X 4 ft. tall simple tables to create a slatted floor, we can configure a mezzanine to hold the chicken feeders, waterers, and nest boxes to protect them from the raucous pigs. Rabbits are suspended in their housing above the pigs as well, to the side of the chicken mezzanine. The chickens can be on the floor or up on the mezzanine. This way we're using the cubic footage of the hoop house and not just the linear square footage on the floor.

When the animals all go out in the spring, we clean out the deep bedding compost and plant vegetables in that dark, rich soil underneath the bedding. That way the hoop houses protect animals for 100 days in the dead of winter and then grow vegetables for 250 days a year. We can still extend both ends of the season by more than a month with the vegetables. Bugs are not a problem because the chickens spent the previous winter debugging things. By the time the bugs do become a problem, it's time for the animals to come back in.

Every structure, every field, every piece of equipment, every

Multi-Enterprise

day, every customer we look at in terms of multiplication, not addition. How can we turn the crank for another sale, another enterprise? What else can we do with this? All of this additional activity puts more hours on the tractor, which spreads that overhead across many more dollars. A few thousand from this enterprise, a few more from that one, and a few more from that one eventually add up not just to an interesting life, but to a functional bank account.

Successful farms are the ones that, as they say, have a lot of things going. I like to see activity, a beehive of people, movement, choreography. That looks most like the landscape we saw on our hypothetical plane ride over 1400 North America. It was abuzz with bison, beavers, and birds. The telltale sign that a farm is in decline is when you drive up and nobody is home, nobody's around, and things seem empty. Successfully profitable farms have an air of business bustle about them, of activity, of getting things done and growing things.

Compare that with walking out into a typical massive corn field in Iowa. No birds. No children. Just corn. Or walk into a typical Tyson chicken house. No people. No children. No birds. No spiders. No bumblebees. Just stench and wretch, with door-to-door chickens for good measure.

In no way can I capture the full application of the multi-enterprise idea here. Chickens in gardens, weeder geese, Indian Runner ducks for insect control, open-canopied nut fruit trees above vegetables, chickens under citrus trees, grape vines under nut trees. The list is endless. No doubt as you head down this path, you will discover synergistic stackable relationships, or guilds, that others have not even discovered.

Nature has no waste streams. In nature, "throw away" doesn't exist. The waste of one thing is the feedstock for the next. Look at your farm like a big never-ending loop and try to eliminate as many dead ends as possible. Some people accuse me of not being true blue because we buy non-GMO grain for our poultry and pigs. "You should be gleaning salvage food from your restaurants or the local supermarket. You should produce your own foodstuffs, your own grain."

197

This is where we need a little realism. You can become so committed to diversity that you end up not being good at anything. You can go bankrupt being over-committed to diversity. I love trees and I love ponds, but you don't see the tree growing in the pond.

Our farm was cultivated to grow grain for 200 years and lost five feet of topsoil. Most of our fields have such shallow soil we couldn't till them if we wanted to. But neighbors, down in lowlands (where all our soil washed), have deep, loamy soil. Remember, one of the reasons for animals is to move fertility that gravitates downhill back up to high ground. A farm has a loop function, but so does a whole community.

Rather than practicing such fierce independence that we do things our land doesn't want to do, or that we hate to do, we believe in mutual interdependence. If we look out past our farms to the greater communal ecosystem, we see a multi-enterprise option much bigger than what any one farm can do. That's the way it should be because not everyone is the same.

We've tried to figure out how to go loaded and come loaded to our restaurants, hauling edibles to them and inedibles back to the farm for the chickens. But for some strange reason a chef, not to mention a food safety inspector, doesn't enjoy walking onto a food delivery truck and seeing barrels of stinking, fly-laden food scraps in there. It just doesn't sit well. Yes, we make compromises. In the perfect world, we'd close that loop. But we don't live in that world. In the perfect world, all those restaurants would have some chickens outside the door and we wouldn't even bring eggs into town.

In the perfect world, we wouldn't have half our income stolen to pay taxes to buy ammunition to arm people to shoot our neighbors in combat. In the perfect world it wouldn't be illegal for us to milk a Guernsey cow and sell the milk to a neighbor. Man, if you get me started on the perfect world, I can go a long time. The harsh reality is that we don't live in a perfect world. To stay in business we come as close to ideal as we can within our context, until such time as the world does become more perfect.

Multi-Enterprise

If I go out of business being stubbornly altruistic what does that prove? That just proves that the world wasn't ready for me. Big woop. What a legacy. In all my multi-enterprising, I have to stay relevant to my context--talents, gifts, interests, resource base.

Remember too that every new enterprise has its own learning curve. Innovation is expensive because by definition it involves a lot of failures. Gaining experience is costly. Nobody can afford to be innovative in every facet of their business. You have to get one enterprise going well, then work on another one. That way the profitable one finances the research and development on the new one. The same goes for changing, or tweaking an ongoing enterprise.

I'd love to raise soldier flies or feed my chickens earthworms. But they are both perishable. When I do a time-motion study on that, at our scale, it doesn't work. Are there things we can do? Absolutely. We've done some pasture cropping and that holds exciting promise. I'm committed to doing some grass silage for the pigs and poultry--not the cows. Just last week Daniel said he had a line on some waste peanuts. Maybe that will pan out. But do the cheap peanuts have more chemicals than our local non-GMO grains? Sometimes gleaning is problematic that way.

We tried gleaning from the food bank a couple of years, but the pork got flabby and we had fields full of plastic and wrappings. It made things messy. The time spent cleaning up and opening bags far eclipsed any savings if we just bought clean, bulk-handled, non-perishable feed from our non-GMO mill. Only you can figure out how many and what kind of enterprises fit for you. That's why you do time-motion studies and gross margin analysis. If it doesn't make economic sense, you have to figure out something else.

I tried to work with a local cheese maker to bring the whey back to feed the pigs. But even at five miles away, the cost of running to get small batches didn't justify the nutrients. Sure, the pigs liked it, but if it doubles my feed costs, suddenly I have to raise my pork prices 30 percent. Talk about hearing squeals--and it won't be from pigs.

Not every tree lives. Not every enterprise will survive. We tried growing pheasants, turkeys, rabbits, and grapes in a netted aviary. The rabbits dug up the grapes. The turkeys ate the bottom leaves. The pheasants ate the top leaves. It seemed cool in theory, but it didn't work. I could keep you up for awhile with things we've tried that didn't work. One year we thought we'd grow earthworms under the rabbits in the hoop houses. We made beds and cloches to go on top so we could sprinkle barley on top of the earthworms and grow fresh sprouts for the chickens by removing one 4 foot section of cloche a day.

It worked great for a month, then it got really cold. The rye quit sprouting, the worms got cold and crawled away to deeper ground, and rats moved into the worm beds. I mean rats by the hundreds. It was like a plague. In the spring we had rats everywhere and had some mass killings. What a mess. The sheer complexity of these systems is both their beauty and difficulty. You have to have the right mix, the right balances, and the right infrastructure. It all has to fit.

You don't see redwoods growing in Appalachia. And you don't see our magnificent white oaks growing in Sequoia National Park. You don't see fish on the land or cows in the ocean. You don't see pineapple in Maine or cranberries in Tahiti. Within the diversification idea is a context. The take away here is to be multi-enterprised, yes, but realize you can't do everything, and you sure can't do everything at once. It doesn't take much multi-enterprising to run circles around any industrial single-commodity farm. The ecological functionality of the average farm in the U.S. is so low, you don't have to have much diversity, in your life, your landscape, or customers, to be way above average.

Be realistic, don't hang yourself on the way to diversity. But realize that most of us, including me, can find additional synergistic enterprises to add to our farms. Not a single farm in the world is fully leveraged. I've not seen a farm yet that couldn't add one more enterprise or do something a little better. That doesn't mean the farmer needs to work more hours; it's simply a statement of productive abundance and profitability. It'll take more people, more management, more creativity, and some time. But it's worth adding enterprises for the multiple benefits they offer.

Chapter 11

Stay Nimble

Traditionally, the least nimble business in the world was a farm. With barns and sheds, silos and wooden fences the stationary nature of farms makes them difficult to retrofit or adjust. All of this stationary infrastructure and the equipment that goes with it is built initially as a time saver but often enslaves the farmer with its use and maintenance.

Imagine an 18 year old farm girl from a confinement dairy attending a grass conference. She drinks everything in and has her epiphany. She gets home late in the afternoon, immediately changes into barn clothes, pulls on her boots, and heads to the milking barn. She's bursting to tell everyone what she's learned. The first person she sees is Grandpa gingerly shuffling around the baby calf shed.

Excitedly, she runs up to him and starts gushing, "Grandpa! You wouldn't believe what I learned today! Can you imagine that we could forget about filling silos and scraping concrete and just turn the cows out and let them eat and poop out in the pasture, sell almost all the equipment, almost eliminate vet bills . . ."

During this verbal explosion, Grandpa, who always has a sparkle in his eyes when interacting with his granddaughter, gradually withdraws, and finally interrupts with palpable abhorrent dread, "Listen here, little missy. Your grandma and I spent a lifetime bending re-bar, begging for bank loans, pouring concrete and putting this outfit together so it would be efficient and now you're just going to, to, to walk away from all this?" With a magnificent sweep of his arm, he indicates all the buildings, silos, and concrete manure lagoon.

With eyes narrow and squinting, unable to hide his hurt, he queries, "Are you really my granddaughter?" Oh, the ultimate put-down: disinheritance. You see, the pride of the average farm--and farmER--is the scale and equity in buildings, infrastructure, and equipment. We become emotionally and economically attached to the things we believe will free us from bondage.

Even if our stationary, capital-intensive, single-use, energy-intensive stuff is obsolete, not fun, or pathogenic, we stay with it because we've invested so much of our self-worth into it. Instead of it freeing or enabling us, it actually ties us down to continued use far beyond reason. The whole idea of nimble business is to maintain exits and exchanges for easy adaptation to new circumstances.

"We've always done it this way" kills innovation. The same is true about stuff because stuff defines how we do things. A nimble farm business is one that's long on ideas and relatively short on stuff. The average farm in America right now requires $4 in depreciable infrastructure to generate $1 in annual gross sales. Here at Polyface, our average is 50 cents to $1. That's an 800 percent difference in the ratio.

The reason the average farmer is now nearly 60 years old is because when young people can't get in, old people can't get out. When the impediments to entry exclude fluid succession between generations, the older generation can't exit. In fact, both generations are stuck. The single biggest hurdle in farming, of course, is land. Nobody wants to build long-term structures and pour concrete on leased land.

Furthermore, many nooks and crannies of unused land exist that could be utilized if infrastructure could be moved from one area to another. How do we lower the capitalization hurdle to enable young people to get in? And how do we take the land capitalization out of the entrance exam? Consider with me three ideas.

1. Mobile farms. What makes a farm a farm is what a farm-ER has done to the landscape. Any piece of land can be a farm if a farm-ER begins doing something to it. A farm is nothing more and nothing less than a piece of land touched by a farm-ER. This is important to understand because as long as you think being a farm-ER requires you to own the land you're farming, this land capitalization problem will plague you at every turn, whether it's to start, to expand, or to change location.

It's what a farm-ER has done to the land that makes it a farm. Well, what is that? Usually it entails growing things, both plants and animals. Those growing things require water, food, habitat, and control. Let's look at these in the case of plants first.

Water is irrigation and hydrology. Using swales, terraces, roof eaves or other features we can create dry spots and wet spots. Food is fertilizer or foliar feeding. For plants, anything that increases soil health I would consider food for the plants. Habitat is a proper sun and climate context. Obviously if you're growing mushrooms your habitat will be far different than if you're growing tomatoes. Weeds, temperature, spacing all have to do with habitat. Finally control is keeping things the way you want them. Pruning, debudding, pinching back cucumber vines, and trellises all fall under the control category.

How can this be mobile? Consider the work of John Jeavons with the broadfork and double digging. This is all done with extremely inexpensive and mobile tools as opposed to big equipment and tractors. Drip irrigation utilizes primarily permeable tape that comes on a roll like toilet paper--bigger, yes, but you get the idea. Raised beds, whether created by a roto-tiller with attachments, a shovel and rake, or a broadfork creates a functional habitat. Intense gardening rather than rows, where every inch is covered with foliage, reduces weeds and yields more production per effort.

Controlling more precisely where seeds fall is part of this idea. The productivity increases when utilization of all these techniques creates location options that would normally be unthinkable. Some of the most impressive work being done in the urban produce field is in British Columbia by Michael Ableman. I can't applaud his work and innovation enough. He literally creates mobile produce farms out of plastic containers that he can move around with forks on a skid steer loader. In short order, he can fill these with soil, load them on a trailer, and go into a vacant lot or old decaying parking venue and set up a working, functional farm.

If you ever have the privilege of visiting Eliot Coleman's farm in Maine, you'll be struck by the lack of heavy equipment lying around. Goodness, go visit Jean-Martin Fortier, whose book *THE MARKET GARDENER* is turning produce operations upside down, and you'll wonder what makes the farm work. He harnesses earthworms to do his tillage. But it all works because his garden plots are identically sized (control) so he can easily roll up and shift an opaque fabric cover from spot to spot. That allows the worms to do their magic. When he removes the tarp, the plot is weed free, covered in earthworm castings, and ready to plant.

I think my best discovery on the plant side of this idea was in Alberta where I met a lady in Edmonton who had a wonderful success story. She lived in a sixth story condominium, had no money, no land, but a burning passion to become a farmer. She knew one person who owned a backyard, so she went to this acquaintance and begged to plant a small garden in the backyard.

"Sure," the owner said. They agreed on a small plot about 12 ft. X 20 ft. in a corner of the backyard. The lady prepared, planted, and maintained the spot, turning it into a beautiful and abundant garden. The lady who lived next door saw the garden, approached her neighbor, and asked one day, "Do you think your gardener friend would be willing to put a garden in my backyard?"

You can see where this is going. By the time I ran into her in Edmonton, she was farming 18 backyards, full time with one part-time employee, and her entire farm infrastructure consisted of hand tools that she carried on a special rack on her bicycle. She literally bicycled from yard to yard, or from farm to farm. What makes a backyard a farm? What a farm-ER does to it. She didn't own any land. She didn't even own a shed or outbuilding. She lived in a sixth-floor condo, for crying out loud, but I'd put her farm bona-fides up against any commercial farmer out there. In fact, chances are she'll outlast many of her more commercial counterparts in farm country.

She doesn't have any equity tied up in machinery, buildings, or land. Talk about nimble. If she loses a backyard, she markets herself to grab another one. In fact, in urban sectors, this opportunity is both viable and gathering some steam. This is the kind of story that warms my heart and should help anyone reading this to understand that no matter your situation, an entry point exists if you keep your creativity high and your capitalization low.

When your infrastructure can move from spot to spot, the size parcels that become viable for farming opportunities drops precipitously. If your tractor has to stay put, the size land base to justify the tractor must be fairly large. But if the tractor services three parcels, each or them can be small but the aggregate justifies the tractor.

Certainly some things are not as conducive to mobile farms, the most obvious being a vineyard or orchard. But the beauty of those operations is that they are conducive to stacking so that the land can be used for more enterprises. Grazing under fruit and nuts with herbivores or poultry, or both, is a great way to add a mobile component to the operation. Often this works best when the stationary trees, bushes, or vines receive greater spacing than normal. Opening up the canopy not only encourages ground cover to grow, whether it's vegetables or pasture for grazing, but also facilitates getting around.

Don't underestimate the synergies in these kinds of things. One of the most mesmerizing things I ever saw was in an olive grove north of San Francisco. One of the biggest costs in olive production is pruning, especially all the little suckers that sprout up from the interior of the tree, from the big trunk and major branches. Of course, mowing is also a big cost.

This outfit had a herd of a couple hundred goats controlled by electric netting. The goats mowed the grass and kept the weeds down as they moved systematically through the olives. That was functional and beautiful but the wow was in how the goats climbed up into the trees. Olive trees are short, stocky, and gnarly. The goats literally jumped up onto the low branches, most of which were nearly horizontal down low, and climbed up as far as they could go. Many goats went as high as 12 feet.

What attracted them into the trees were all those nasty little succulent suckers, which the goats dispatched with fervor and finesse. Mesmerized, I could have watched this theater for hours. The electric netting surrounded about an acre and the farmers moved it about every second or third day, depending on density of the ground cover. Half the goats at any one time were on the ground, but the other half were all up in the trees. Certainly one of the reasons everything appeared so elegant was that these guys had learned how to handle goats, how to set up the netting so it functioned with good spark, and a host of other nuances.

They could easily have offered mowing and pruning services to other olive growers in the region. The whole service was completely mobile. In California, around urban areas, more and more opportunities exist for mobile grazing operations. The risk of spark from a steel mower blade hitting a rock and igniting a wildfire is making mechanical vegetation suppression obsolete. The new alternative is grazing.

The skill level, however, required to move a flock of 400 or 500 sheep and goats onto the edge of an expressway, keep them under control, and then move them on is tremendous. But these outfits use ancient shepherding skills coupled with modern mobile infrastructure to create completely mobile farm operations in the urban sector. Brittany Cole Bush, who dubs herself the Urban Shepherdess, is trying to start a school to teach these skills.

Coming closer to home here on our farm, our love affair with the mobile farm dates way back to our earliest days. Dad was a fan of mobile infrastructure, partly because it freed us up from building something we regretted later, either from a design standpoint or a location. Once you build something stationary, you're locked in.

While you can theoretically move a structure, practically speaking unless it's on tires or skids, you don't. Keeping the infrastructure mobile frees you up from being locked in. Furthermore, mobility by definition requires simpler design.

Perhaps nothing has enabled this to happen more than modern band sawmills. Sawmills up until the 1980s all used a circular blade that weighed 800 pounds and took a kerf of a quarter inch. The kerf is the amount of wood removed so that the blade can go through. These big thick circular blades removed a lot of wood. Not only did that require a lot of horsepower, it also meant the loss of lumber.

A quarter-inch kerf turns an inch of lumber into sawdust every four cuts. The reason early Americans built log cabins and post and beam construction was because milling small dimensional lumber was too expensive. Without chainsaws and efficient milling, it was much easier to use large dimension wood than it was to cut it down to a smaller size.

Today, modern band sawmills take a kerf about a tenth of an inch. That means in ten cuts we remove the amount of wood that an old circular mill removed in just four cuts. Since we're not turning as much wood into sawdust, we can power that band saw with a much smaller engine. Whereas the old circular mills required 100 horsepower, today's band sawmills require only 20 horsepower and these little engines can run all day on a couple gallons of fuel.

The big breakthrough was the day someone had the idea that instead of holding a 3,000 pound log within a thirty-second of an inch tolerance and running it across a stationary blade, why not let the log sit there stationary and run a lightweight horizontal band saw through it? The old mills required several tons of pig iron to accomplish what these mobile units today do with a few hundred pounds. This means we can literally cut tinker toys efficiently and build infrastructure out of lightweight, small dimension lumber. Whether on skids or rubber ties, these new lightweight structures can be loaded on trailers and moved from location to location--the ultimate portable farm.

Portable structures can be moved daily on a given location to keep the animals healthy and eliminate manure hauling and other heavy maintenance required of large scale stationary infrastructure. A further advantage is that they are neither classified as equipment or buildings. That means they do not require building permits and are not registered as assets, so no personal property taxes or real estate taxes. The government has no clue how to classify these things. Is that not the coolest thing in the world? I love farming with infrastructure that is invisible to government agents.

Because they aren't machines, you can expense them as you build them. No depreciation. Just expense everything day one. It's like the ultimate Stealth farm, invisible to tax and governmental radar. Even though we're a farm cranking a couple million dollars in sales, we have nothing to show for it. No confinement factory houses, no big silos, no big tractors. The whole outfit is invisible.

May I suggest that's one of the neatest things about the mobile farm? It can be imbedded in the ecological nest. When you think about modern orthodox farming, all of its activities and manifestations dominate the landscape. Whether it's the massive feedlot with its fecal pall hanging like a nuclear blast cloud over the landscape or crop dusting airplanes nuking fields occupied by hazard-material-suit-clad workers, everything about the industrial orthodox farmscape brings attention to human domination. It's a "look at what I've built, what I've done." It's a hubris celebration on steroids.

But the mobile farm celebrates lightweight, gentle, almost invisible infrastructure. Cars whizzing past wouldn't even know we had hundreds of pigs in the silvo-pasture adjacent to the highway. Not only would motorists not see them, they couldn't even smell them. I call this nook and cranny farming because the mobile infrastructure can be tucked anywhere and doesn't obscure or dominate the landscape.

Electric fence offers mobile control of animals unlike anything available in human history. We can hold thousands of herbivores in a spot with nothing more than a thread carrying an electric shock. With computer micro-chips in energizers, we can increase the volts at extremely low amperage. Not long ago, electric netting that weighs 12 pounds per 50 yards and can keep poultry in while keeping bears, coyotes, and raccoons out would have been unthinkable. When I was a child, the old mechanical electric fence energizers sent pulses a quarter of a second in duration. That long spark time created a lot of resistance and melted polyethylene.

But today, with spark lengths down to one 250-th of a second, the duration isn't long enough to create heat. As a result, metal conductivity threads can be woven with strong polyethylene fibers to create an electrified grid that is strong, lightweight, and highly effective. For the first time in human history, we can grow commercial-scale flocks of poultry, for example, in a more bird-friendly, hygienically, and environmentally-enhancing way than you could a backyard flock on a 1920 farmstead. Is that not the coolest thing since sliced bread?

Kevin Kelly, author of *NEW RULES FOR THE NEW ECONOMY*, pointed out that everything is being miniaturized, downsized and restructured. He said the 120-pound secretary has been replaced by a 4 oz. mail router. Think about the modern car. Pig iron (mass) is being replaced by information technology, which weighs nothing. Today's car knows ten times as much as the 1960s muscle cars, but weighs half as much. New lightweight super-strong alloys and micro-chips are fundamentally changing the size and weight of everything.

Applying this to the farm helps us to understand why the factory farm is obsolete; the industry just doesn't know it yet. Since food and farming are the oldest professions in the world, they necessarily are the slowest to change. Farmers tend to be conservative. As media and manufacturing go toward robotics and lightweight everything, industrial farming is still trying to build more concrete, bigger buildings, and more weight. While everything else is using swarm design, industrial farming is still trying to cram more stuff in a single location. While the rest of society moves to the cloud, agriculture pours more concrete.

Mobile lightweight farming infrastructure more approximates technological advancement than any other agricultural design idea. Think about what woven polyethylene now offers. If you wanted sun protection for cattle, turkeys, or pigs not very long ago, you thought about stationary trees or sheds. If you thought about mobile shade, the problem was wind on metal roofing. Opaque roofing blocked the sun alright, but it also caught the wind like a sail. Anything light enough to move under that opaque roof would blow away in the next heavy wind.

Plastic shade cloth came to the rescue. It's now used on ginseng farms, in horticulture, but more and more frequently for portable livestock and poultry shade. Available in all sorts of sun blockage (40 percent to 80 percent), it's a mesh and allows the wind to go through. We use it for pigs, turkeys, cattle, sheep, and waterfowl--anything that doesn't mind getting wet. A 100 mph wind comes roaring across the fields and the shade cloth just sits there atop a tinker-toy skeleton rippling gently. The lightweight structure on tires doesn't even wiggle. Amazing.

Attached with bungee chords or baler twine, the shade cloth can be loosened and folded up like a big bed sheet. The roof purlins can be taken down and tied onto the hitch. Within a few minutes we can go from functional to broken-down, hook up to a pickup, and go 70 miles an hour down the interstate to another location. Did someone say something about a portable farm? Have farm, will travel.

We've made portable chick brooders. All you need is a mobile home axle. Polyface currently has some 200 tires involved with all our portable infrastructure. We're fiends for axles. Anytime an auction occurs, we buy used axles. They're cheap. With the band sawmill, we can cut out heavy stringers, build a wooden chassis, bolt on a nice hitch, and we're up and running. We've built a couple of 8 ft. X 16 ft. portable chick brooders big enough to start about 600 chicks. We divide that space into three sections of 200 chicks apiece. These brooder houses tow just like any other trailer.

We call the turkey shelters Gobbledygos. Simple V-trussed structures, we put roosts across the V-trusses and the whole arrangement weighs only a few hundred pounds. It's like a glorified wheelbarrow with a shade cloth 16 ft. X 30 ft.--big enough for 400 turkeys. A feed trailer hooked to it creates a two-car train that we can tow around anywhere we want. Highly efficient when surrounded by the electrified poultry netting, turkeys get the benefit of outdoor, fresh pasture every day, and we have a completely unidentifiable, invisible, low-capital, expensable farm.

What's not to love? The Millenium Feathernet is on skids using an X-truss arrangement to support an A-frame. By the way, all this poultry stuff is in my book *PASTURED POULTRY PROFITS*. I have no intention of duplicating all of that information here. The point I'm making is the functionality and nimbleness of portable infrastructure. The less you nail down, the more options you have. As soon as you start nailing down, you're obligated to the location and the design. All of our infrastructure is light enough and cheap enough that even if it rots in a decade, we're still money way ahead. And we didn't have to ask permission to build any of it. Now that's real freedom. Instead of "it's a bird. It's a plane. It's Superman!" We have "It's a building. It's a machine. It's Mobile Farm!"

All sorts of permutations on this theme exist. Anyone familiar with Eliot Coleman's three-stage hoop houses can appreciate what mobility can offer. He has these on a track and pushes them from one spot to another. That way he can not only rotate hoop house spots, but can get something started and then move the protective cover to another space to grow something else while the started vegetables actually grows better without any cover.

We've built hoop houses for hay storage or winter chicken housing on rental farms and then picked them up and moved them if the land was sold or we didn't renew the lease. This is what I call quasi-mobile. When a structure is simple enough that in a few hours you can dismantle it, throw all the pieces on a flatbed trailer, take it up the road and reassemble it somewhere else, that's not truly mobile but it's close enough. The Mongolian yurt comes to mind. It's not a trailer, but it offers a lot more options than being permanent.

Before leaving the mobile discussion, let me point out one more advantage of direct marketing: you can move and your customers will stay with you. You can't move but so far, to be sure,

Stay Nimble

but within about 50 miles you can move around and your customers don't really care whether they have to drive north or south to get to you. You can take your customers with you when you travel. But in industrial farming, a 50-mile move can often take you out of a service or processing area. This is yet another reason to look at direct marketing.

The ultimate point of all this, of course, is that mobility frees us from being tied to a specific piece of land. When we cut that shackle, we create lots of opportunities. Now the second aspect to being a nimble farm. Be forewarned--it's another M.

2. Modular. If you decide to become a Tyson chicken farmer, what's the first thing you have to do? You got it, build a half million dollar factory chicken house. I would call that a slight impediment to entry, wouldn't you?

Isn't it odd that the industrial folks always hurl the epithet of "elitist" at me because my prices are higher than supermarket junk, but they don't consider the fact that only people with big mortgages or deep pockets (or no sense! ha!) can get into their systems? If we're going to preserve ease of entry, we need a model in which anyone may enter, preferably with pocket change.

I'll use our pastured broiler set-up as the best example. You can get all the details in my book *PASTURED POULTRY PROFITS*. With these little mobile, modular, floorless shelters, each one is cheap enough that all you have to do is not go the movies for a couple of months and you can build one. It's only a couple hundred dollars. If you like it, you can build another one. If you don't like it, you're not out much. Each one is only 12 ft. X 10 ft. X 2 ft. high, or 120 square feet, and houses 75 broilers. We move them every single morning to a new spot.

If you're crazy nutso in love with it, you can be like us and build 150 of them. That's equivalent to 18,000 square feet, or a house 90 ft. X 200 ft. That's a fairly substantial house, but because this is 150 modules, we didn't have to bite off the whole enchilada the first time.

We built a handful and used them for a couple of years. Then we expanded and built a few more. The 150 we now have represent the current state after a few decades of production. We can throw 8 of them on edge on a 24-ft. gooseneck flat bed trailer and move them from farm to farm (or more precisely, move the farm from location to location). Man, it's hard to change our terminology, isn't it? It's kind of like saying "I'm going to church." No I'm not. I'm the church on the way to meet. The building is just a building. But I digress.

The beauty of modular infrastructure is that it allows us to enter small and then scale as big as we want without prejudice. One of the critical elements of the industrial model is that it scales up extremely well, but does not scale down well. Anytime we can create a model that scales either direction without discrimination, we know we're onto something. In my book *THE MARVELOUS PIGNESS OF PIGS* I call this a "whosoever will" idea. Social scientists might call it the ultimate democratization prototype.

I don't care what you call it, but I know that models that carry no prejudice regarding size are the most nimble. If a backyarder wants to jump in, great. If someone with sights set on building a fairly large business wants to move to that level, that's fine too. And if you're big and want to scale back, you can easily decommission modules without major financial jeopardy. How do you scale back a Tyson chicken house? It only works at a certain scale. But with modules, each one is a self-contained, functioning entity.

Modules allow you to respect size balance. Sticking with the broilers, for example, they have a numeric sweet spot. If you have more than 100 three-week old chicks out on pasture and the temperature drops overnight to 20 degrees F, you'll pick up a couple of dead ones. What happens is that as the night cools off, the chicks huddle together for warmth, crowding into each other tighter and tighter. With more than 100 birds in the clump, the pressure on the couple of middle ones is too great and they actually suffocate. They don't get too cold; the others crowd around and on top of them, snuggling them into suffocation.

From an animal welfare and health standpoint, therefore, it's much better to have several groups of 75 chicks under those conditions than it is one group of 500. Goodness, in a group of 500 you'll probably have 20 suffocate. But in a group of 75, the mass is not big enough to suffocate anyone even though pressed tightly together. Chicks are actually much hardier than most books will admit, but all these book writers are used to large groups of 1,000 or more. As the group size increases, anything that pushes them, whether it's weather or fright, is multiplied once you go over a certain number.

My unscientific, anecdotal, experiential stress equation is this: mass X density X time = stress. You can have a couple of animals fairly tightly contained and they'll be fine. But put a big group (mass) at that density, and they'll be stressed. By the same token, you can have animals fairly confined, like in a corral, but as long as they aren't in there long, they're fine. As soon as you leave them for an extended period (time) then they morph into stress.

If we have a batch of 1,000 broilers, therefore, it's much healthier and more productive to have them split up into many different groups than have them all in one big group. This is true on many fronts. From *SQUARE FOOT GARDENING* to commercial orchards, the idea of beds, blocks, and modules simplifies organization. Many years ago when we converted the garden from one big area to raised beds with paths in between it immediately became more child friendly. With one big area "get out of the tomatoes!" too often made our children hate the garden. But as soon as we converted to beds, the walkways for people and the dedicated space for plants were easy to see, even for little children. With clear lines of demarcation between paths and active growing space, modular design made the garden much more child friendly.

Modular gardening and farming has psychological benefits. You can measure progress much easier. A bed of green beans 20 ft. X 3 ft. may contain 60 feet of linear row, but it doesn't look as long. When Michael Ableman moves his garden modules onto a derelict city lot in Vancouver, each one can be prepared, planted, and picked in a matter of minutes. If he has 1,000 of these on a site, it might represent, in aggregate, an extremely large vegetable operation. But the break-up in modules gives it a small garden feel.

Modular design encourages better management because rotations and segmentation can be delineated easier. You can visualize rotations. Even drawing out the production plan on paper is easier with modular growing spots. You can designate each module with a number or letter. Workers can easily identify which area is which. If it's just kind of one big mass, everyone is confused and the whole thing breaks down.

David Schaefer, founder and innovator of Featherman Poultry Processing Equipment, has developed what he calls a Plant-In-A-Box, or PIB. These are federal inspected poultry abattoirs built into empty shipping containers. He builds it out to your specifications, throws a set of wheels under it, drives it to your place, sets it on four concrete pillars, pulls out the wheels, and it's ready to go. The modular shipping container conversion industry is exploding right now with tiny homes, root cellars and a host of uses. Since America doesn't manufacture anything anymore, our shipping container bone piles are growing by thousands a month. This is a real opportunity for anyone looking at modular design.

The nimble farm, then, is mobile, modular, and now the third M:

3. Management intensive. This is where the industrial folks point their finger at me and yell, "Gotcha! I knew it would take more labor." The accusation against the kind of farming I practice is that it takes more people. It's similar to the accusation that cows are inefficient. Well duh, of course they are, which is why they leave enough behind to grow soil and increase fertility. That's why the deepest and most fertile soils on the planet developed under prairies and herbivores, not trees, bushes, and squash.

Rather than deny the additional labor accusation, I embrace it. Of course our system takes more labor . . . ON THE FARM. That's far different than more labor overall. The industrial orthodoxy, or factory farming, takes labor away from the farm and compensates with energy intensity, capital intensity, pharmaceuticals, chemicals, and environmental remediation projects, not to mention doctors, counselors, and lawyers.

The system I advocate puts people back on farms. I don't apologize for that. Anyone reading this book is welcome to come by Polyface anytime and check out my office. Let me tell you, friends, the worst day here on the farm is better than the best day fighting traffic, confined in a cubicle, and attached to electronic devices. From my experience, millions of people would gladly change their circumstances IF they thought they could make a decent living at it. That's what this book is all about.

Yes, we strategically and purposefully replace pharmaceuticals, capital intensive infrastructure, energy intensive infrastructure, and a host of experts with actual warm bodies right here on the farm. If you're not going to douse everything in chemicals and drugs, it takes more observation and management skill to keep things healthy. Wes Jackson calls this "eyes to acre ratio." I like that. Wendell Berry explains that in order to offer good care, you need to know what you're caring about. It's easier to care for a few acres than a million acres.

The notion that we can provide good land care with driverless tractors following GPS guidance systems managed by assumption-based scientists a thousand miles away is inherently foolish. In many ways, land care is a lot like parenting, and nobody would defend parents who speak to their children from a distance through some sort of R2D2 device. Stewarding done well requires intimacy; intimacy requires proximity; proximity requires personal acquaintance. All the elements follow like your hand in a glove. You can't parent from a distance; neither can you care for land from a distance.

Customization is the art of personal identity. How can a USDA nutritionist formulate dietary guidelines with no understanding of my microbiome? The one-size-fits-all mentality extended into the fabric of our streams and hillsides is the problem, not the solution. Yes, here at Polyface we're a farm with people. When you drive up to the front gate and honk the horn, people will pour out of buildings, fields, and houses like a disturbed ant colony. We're not a place where plants and animals grow without human interest. The biology under our stewardship enjoys a lot of personal attention; that's how we keep the drugs away, the bankers away, the hordes who make their living extracting rural wealth and rural labor from a colonial serfdom.

I've had numerous Wall Street investment outfits call me, "We want to put money in sustainable agriculture. Can you help us figure out where to put it?" Every one of them is looking for a piece of equipment, a small business to grow into an empire. Of course, they're all looking for the service or product that will give them 10 percent on their money.

One guy hung with me for nearly two years. Great hearted fellow, truly. We exchanged many emails, talked on the phone. He even arranged a high powered bus trip from New York with a bunch of investment fund movers and shakers. But in the end, it, like all the rest, collapsed because none of them could figure out how to make a bunch of money off me. See, we're not putting money in things; we're putting money in people. Our big budget item is not tractors and buildings and petroleum and chemicals; it's people pay in commissions and salaries.

People are the beneficiary and recipient of this farm. What if we had a food system that emptied half the hospitals, half the prisons, half the drug companies, and half the food regulators and instead offered substantially more meaningful work out in the fresh air and sunshine? If you attend any business conference today, you'll hear about two fears over and over. The first is that we have a woefully uneducated work force. We're sliding backwards academically. Our graduates don't know how to write, how to speak, how to think analytically.

The second problem is what to do with all the vocationally-oriented young people. Believe it or not, everyone doesn't want to sit in a Dilbert cubicle at the end of an expressway working for "the man." A large segment of the population actually enjoys getting their hands dirty, doing craft work. Many people think our culture will collapse at some point; historically speaking, that's the way to bet. Regardless of whether unsettled times come due to sickness, economic malfunction, energy snafus, or exterior invasion, you want to be friends with people who know how to grow things, build things, and fix things.

Growing, repairing, building--that's the foundation of economy. I suggest that people on farms, doing noble, sacred work like healing gullies, building soil, producing nutritious non-pathogenic food, can provide affirmation and meaning to those of us who don't thrive in a cubicle. Do we need folks running high tech stuff? Yes. And some people thrive on it. But a lot don't. They don't want to go to college. They love working with their hands, wearing work clothes, and sporting calloused hands. I've just described me. Even now, at this stage in this book writing project, I'm chafing to get out there on the sawmill and get some real work done. I've been sitting in front of this computer too long. Time to go make some chips fly and smell sawdust. Nothing is as intoxicating . . . or affirming.

This management intensity that I'm describing is highly skilled. This is one reason why here at Polyface we don't do short-term internships or two week pop-ins. Good management is highly skilled. Looking over a flock of chickens and noticing everything quickly, actually seeing all the flock nuances, only comes after years of experience. Just the other day Daniel and I and two apprentices were going over to one of the rental farms to do some work.

I got my chainsaw and other tools together while Daniel went up to the hoop house to get the two apprentices. They were up there putting some hay in for the pigs. They'd been up there for half an hour working, so he was sure they were about done. Suddenly he came running back all out of breath to get some plumbing fittings. It was Christmas and one of the other apprentices had left for home on Christmas morning. For a week the temperature had been dropping well below freezing at night so we'd been turning off the water and draining the hoses in the evening.

Christmas Eve clouds rolled in and the temperature rose so Daniel told everyone to leave things hooked up overnight since it was not going to freeze. That would also take off another chore on Christmas morning. Nobody checked that particular hoop house Christmas day--we had lots going on with gifts and meals and parties. So here we were, the following morning, and it turned out that the apprentice who went home on Christmas, following the previous week's routine, had turned off the water on Christmas eve before he went home. So the pigs drained their waterers Christmas day and by the next morning were thirsty.

That's all background. The critical thing to know is that when the other two apprentices went up to feed the hay the morning after Christmas, the pigs had dumped over their empty waterers and torn up plumbing fittings as a result. The apprentices had been up there, in all three pig pens, putting in hay for half an hour, next to torn up waterers, and NEVER NOTICED! They were literally bumping into the overturned waterers but it never dawned on them that they were overturned. Folks, I could keep you up all day with stories like this. The point is that when I say management intensive, I mean management intensive. It's not just slow down in the pickup and scan the herd.

It's not just zip by in the ATV. It's not walking by like a zombie, eyes glazed over and mind in la-la land. It's full on, full engagement, all senses at red alert, all the time. Not some of the time. All the time. Frankly, it's a lot easier for farmers to just give a shot of drugs. It's a lot easier to borrow money for infrastructure that allows brainless people to do the work. In a way this whole management intensity is our firewall of protection from competition. Very few people have the stamina and situational awareness to develop the skill that will enable them to be successful in this kind of model. Anyone CAN do it, but most WON'T do it.

The real equity on the farm, then, becomes skill, information, and customers. No bank or loan agency can ever call you up and foreclose on your skill. Nobody will ever file papers against you to repossess your information. Nobody will bring a tow truck and haul your customers away. When equity moves from depreciable physical structure to nonphysical skill and information, it fundamentally changes the impediments to entry as well as the risk of financial exposure.

When we talk about a nimble business model, it's much easier to adjust our thinking to a new piece of information or something we've observed than it is to retrofit a million dollar combine or a half million dollar Tyson chicken house. And it's a lot safer to have knowledge and skill as our equity than that mortgage paper. Just to tie up the loose ends, remember that we don't have to pay taxes on skill, information, or customers. We can have as much of those things as we want and it never goes on an IRS form. Isn't that the coolest thing since grits?

When farm equity is in physical structure and the farm runs more on petroleum and pharmaceuticals than our own observational skill, we're always subject to breakdowns and commodity price fluctuations. Often these are not things we can do anything about. Even if you think you can stop breakdowns with preventive maintenance, you still have the cost of that preventive maintenance. If war breaks out and petroleum doubles in price, you can't do anything about that. The sheer dependency on things outside our control creates a fragility to the farm's existence.

But when our equity is nonphysical, unless we lose our minds or destroy our character, the things we know and know how to do are immune from these vagaries. That's a healthy place to be.

The single biggest downside of what I've just described is that all of our tax and regulatory law favors infrastructure and drugs rather than people. You can abuse your tractor and write off the repairs on your taxes as expenses. If you have to replace the tractor, you can depreciate it. If you buy drugs, you can expense them.

But if you hire someone, you have workmen's comp, withholding, and all sorts of labor police from minimum wage to occupational safety crawling around looking for frayed extension cords and 15 minutes of unpaid service. Tax law and official government policy rewards NOT having people. As America goes more socialistic, this will simply increase. Don't interpret this as condoning abusing people. I'm certainly not suggesting a person is no different than a tractor. I'm simply pointing out the difference in attitude and risk, and of course this is one reason why businesses go to great lengths to not hire anyone.

Certainly the fiefdom subcontractor concept helps some--and some would accuse me of shirking my social justice duties for pointing this out. Until you've been sued or gone through a workmen's comp audit, keep your mouth shut. Politicians start all these programs in good conscience, but in the end they actually reduce opportunity and destroy the nation. Workmen's comp, for example, is highly prejudicial against my kind of farming.

Let's take an exposure actuarial like a poultry worker. If we sign up someone as a poultry worker, the actuaries are based on industrial contexts. That's a Tyson chicken house full of fecal particulate (not good for breathing, to be sure), with lots of whirring fans, feed augers, motors, electrical conduit. It's a pretty unsafe place. The state-mandated actuaries reflect this. But on our farm, a poultry worker simply goes out to the field, moves some light shelters, carries some water and feed buckets around, and walks home. No electricity, no motor, no fecal particulate, no feed augers. But we pay the actuarial based on industry risk. So when the industry uses depreciable, energy-intensive (often subsidized) and deductible expenses rather than people, the system rewards a non-person poultry operation.

But if we don't use any of that highly expendable and depreciable stuff, substituting people instead, we not only pay an unfair amount into a system prejudiced against us, but also have a highly prejudicial substitution situation. The people we hire to replace all this expendable and depreciable stuff carry a host of overheads and regulatory burden that the other expenses did not. It's just a point that needs to be mentioned. It is what it is and I still do it, but I think we need to recognize that all this wonderful-sounding warm fuzzy verbiage about people-centricity rather than mechanical-centricity carries a price tag. Sometimes that price tag is buried in unusual places.

For example, through minimum wage laws the government decides the price tag on a worker. But machinery and materials are unencumbered and on the open market. If I can get a tractor cheaper, great. If I can find lumber cheaper, great. But I can't exercise the same business freedom on labor because that comes with a mandated price tag. The result is that if I choose to replace machinery with manpower, our regulatory climate prejudices against that decision. The result is that I'm incentivized, as a business, to always figure out how to get rid of people and favor things over which I have more decision-making freedom. I'm sure you socialists are ready to burn this book, but I think when we talk about business, we have to own up to the real consequences and pressures of our decisions and protocols. This is not to throw a damper on what I've described as a nimble farm. It is to strike a balance and bring fairness into the discussion, as well as to recognize it's not completely one-sided.

The nimble farm can be placed anywhere, at any scale, with equity that is immune to foreclosure. Don't you think that's cool? I do. The nimble farm is mobile, modular, and management intensive. In theory and practice, designing with these principles in mind will put us way ahead of the competition and create stability in an otherwise uncertain vocation. This is a way in for newbies, a way out for oldies, and the secret to functional partnerships, building additional enterprises on land that is owned, borrowed, leased, or squatted. That's an exciting prospect.

Chapter 12

Time and Motion Studies

Product and project flow occupy huge amounts of analysis time in businesses. Way too many farmers come to the vocation as a lifestyle or a way to change the world, or a means to connect with the earth without appreciating the hard business realities ahead. Yes, it's all those warm fuzzy things, but it's also about knowing what to do and doing it efficiently.

Several years ago McDonald's sent a representative to see me about the possibility of offering a grass-finished beef burger in their restaurants. This fellow was an attorney from Washington D.C. who was on retainer for them. We had a delightful discussion and one of the things that came up concerned their test restaurant. According to him, they have a test site and crew that spends every day reconfiguring furniture and layout to find new efficiencies. He told me that if they can trim half a second off a routine procedure, it's worth $10 million to the company. Folks, that is a benchmark of refinement we'd do well to appreciate.

We've got to get our act together because lots of really sharp people are out there trimming minutes and procedures. When we first began processing chickens for sale here at the farm, we set up our disassembly line: kill, scald, pick, go over the carcass for pinfeathers, then eviscerate. It pains me to say it. We figured the evisceration was the dirtiest part, so put it at the end, right?

Wrong. What you want to do is go from whole to finished in a progression of biggest to smallest job. And the longer the viscera stays in the bird, the longer the carcass stays warm--think bacteria growth--and the stiffer everything gets. A friend who was a time and motion engineer in manufacturing was standing there watching us one day and nonchalantly asked, "Why don't you do the pinfeathers--quality control--AFTER evisceration?"

We hadn't really thought it through much--it was just the way we'd always done it. To humor him, we switched stations. Immediately we increased our speed by 30 percent. As it turned out, many of the pinfeathers around the vent and tail that were the most troublesome left the carcass with the viscera. By the time we cut out the oil sack on the tail and around the vent to get the viscera out, half those little pesky feathers we'd been slaving over didn't even exist any more. What a breakthrough.

I was visiting a farm in California several years ago and you know how steep much of that land is. The farmstead was down in a nice valley and steep pastures radiated out all around. They had bought into the pastured poultry concept but were frustrated at the effort required to pull those chicken shelters up the steep hills. I went out to take a look at things and asked a simple question, "Why don't you start at the top of the hill and move them down every day?"

You could see the wheels turning and the light bulbs going on. They had never thought of it before. Why? Because the chick brooder was at the farmstead, which was at the base of the hill. Without thinking it through, they placed the portable shelters near the brooder to start--thinking about the efficiency of bringing out the chicks. By saving a few minutes on the brooder-to-field chore, they created five weeks of uphill struggle. As soon as they took the chicks up to the top of the hill to start the pastured process, it fundamentally changed their daily chore time.

But lest you think I'm real clever, I'll tell another one on me. We run our pastured broilers in two large rectangular fields. The long way is east and west. Our farmstead is on the east. Because our winds come from the west, we always run the fully capped end to the west and the open end toward the east. That also lets the early morning sun in, and chickens sure like the early morning sun. The fully capped end is a bit dark and chickens don't like going toward the dark; they like going toward the sun.

The two fields lie parallel to each other with a strip of woods in between. Our only access to both is on the east end. You can't drive from west end to west end. We cover about a field and a half every year with the birds, so partway through the season we pick all the shelters up, put them on hay wagons and trailers, and cart them over to the other field. Obviously if we start on the west end and finish on the east, we relocate them to the east end of the other parallel field and move them west for the rest of the season.

We'd been doing this routine for many, many years. We knew they were harder to move west than east but oh well, that's just the way it was. Neither field was big enough to accommodate the whole season's moves, so every year at about two-thirds of the way through the season we had this relocation day to the other field. Since this was always done on the east end, part of the year they were easier to move (heading east) and the other part a little harder (heading west). Get over it; it's the way it is. Until an intern one year dared to ask, "Why?"

Well, because our access is on the east end of the field so that's where we set the shelters. It was essentially the same issue as the farm in California, except ours wasn't a terrain issue; it was simply a direction issue. But it had a lot to do with how the light, which draws the birds, entered the shelters. He pointed out that once we had the shelters loaded on a wagon, carting them to the west end of the adjacent field and moving them east added a few minutes to the big mid-season shuffle, but by saving ten seconds per shelter per day, at 60 shelters, that added up to 600 seconds, or ten minutes per day. In 80 days, that's 800 minutes, which is more than 13 hours. Two days of labor.

But beyond the time issue, the birds had much less likelihood of getting caught by the trailing shelter edge when it moved east because they crowded toward the light. They could see the grasshoppers and crickets ahead and didn't loiter back in that dark cap on the west end. Such a small change, but it accumulated measurable efficiencies over the course of the season.

What you learn pretty quickly in time and motion studies is how little things add up. For example, in his outstanding book *THE LEAN FARM*, Ben Hartman explains that on their $100,000-plus gross income 1.5-acre farm, they built tool sheds and bought duplicate hand tools. Although these obviously had a cost, the time savings in always being within a few yards of the tools justified the investment. Walking that extra 50 yards back to a central shed all the time for tools added up over a season. Even a 1.5 acre farm benefited from strategically placed modular tool sheds as opposed to centralized storage.

Especially for repetitive work, which a lot of farming is, little changes can yield big efficiencies. One of my favorites is teaching interns how to dig a post hole and install the post. Economies of motion abound. First of all, buy a heavy duty post hole digger. And no, they don't exist at your local hardware store. You need to go to A.M. Leonard or someone like them to get the real heavy duty tools. The reason you want heavy tools is because when the tool is heavy, it does a lot of the work. When you bring a heavy post hole digger down on the ground, it has some inertia. A lightweight one stops at the first pebble. You want something that will have some oomph behind it.

Grip the handles together when you're dropping the digger. Don't grip them as individual handles. Clamp them together and hold them as one; that keeps the jaws from being slightly closed and it focuses your arm energy straight through the tool instead of siphoning it off in different directions. Don't take short choppy strokes, but lift the digger high and take deliberate, systematic strokes. A few well-placed and aggressive strokes will do a lot more with less effort than a bunch of short, choppy jabs. Never close the jaws until you have a mouth full of dirt. Closing the jaws on half-full loads of dirt just wastes time and energy. Take your time to work with the soil so when you do close the jaws, you're lifting out a full load.

Make the hole just big enough so you can place the post against one side and tamp the other. That way you don't have to remove nearly as much dirt. Make sure you keep the sides absolutely straight so you can pry against them as you go down. If you have tapered sides. you can't pry against the side immediately above your digging. Prying against the rim of the hole gives you no leverage. If the digging gets difficult at all with the post hole digger, loosen the soil with the wedge end of a digging bar (a digging bar has a wedge on one end and round tamping head on the other end).

Always place your dirt pile uphill of the hole so when you refill, you're scraping downhill. Once you have the hole dug, place the post in it and put just a about six inches of fill in the bottom and tamp it. Again, tamping needs to be done aggressively. This is not the time to go at it half-heartedly. If you want that post solid, pick the bar up high and bring it down. Again, fewer strokes, but strategically placed and aggressively propelled will get you farther than a bunch of short choppy strokes.

Once the post stands on its own, position yourself so you're always scraping the fill dirt toward the hole. The most common mistake I see novices make is shoveling into the pile and pushing dirt away from the hole. You want to bring the dirt to the hole, not push it away. And you don't want to lift the dirt; you want to use the shovel's edge, vertically, like a dozer blade, to pull the fill pile over rather than picking it up. Position your hands on the shovel handle so you're pulling it toward you, not pushing it. You have much more strength pulling than pushing. That means positioning the hand toward the shovel with your fingers up rather than down. That means you can use your big biceps to pull the dirt toward the hole.

Always save the sod, roots, and rocks aside as you're filling and tamping. Fill and tamp, till and tamp. When you get it tamped about level with the surrounding ground, then put the rocks, roots, and finally the sod on top to mound up the dirt around the base of the post. That ducts water away from the post to reduce rot. A couple of years ago I was with one of our interns, who happened to have a masters degree in mathematics--she was a sharp cookie. After coaching her through this procedure as we installed a fence post, she stepped back and sighed, "Wow, I had no idea digging a post hole and setting a post was that technical and sophisticated."

I grinned and replied, "Welcome to farming. Everything is like that; you just don't know it yet. The only reason I make it look easy is because I've done it for decades. True skill is all about making something difficult look extremely easy. Never underestimate the hidden nuances of time and motion."

I see this routinely on the hay wagon. When we're baling small squares, we pull a wagon behind the baler and stack the wagon as we go. Not only is stacking a wagon a skill but just handling the bales can be done arduously (clumsily) or gently. The difference is about 50 percent of the energy expended. Using your body and your core to move the bales rather than your arms comes naturally to someone who has done it for years. It doesn't come naturally to someone who spends all day sitting in front of computer screens. Bouncing the bale on your thigh as you lift it up decreases effort by half.

George Henderson in *FARMER'S PROGRESS*, which goes a step farther than his iconic FARMING LADDER, spends a great deal of time explaining how to carry a bucket. Carry a bucket? Isn't that just no-brainer work? Actually, it's not. Body position, hand position, what to do with your thumbs. It's a beautiful art form and anyone who thinks it doesn't make any difference how you pick up a bucket has obviously not picked up very many buckets.

Eliot Coleman is nuts about time and motion, and if you ever attend one of his seminars or better yet go visit his farm in Maine, you'll see the intersection of beauty and function. Every square inch has purpose and every motion is directed. What looks like just work actually is a carefully contrived series of bodily motions. We farmers can't afford to waste energy.

The reason I can hand eviscerate a chicken every 30 seconds for hours on end is because I don't dink around with each bird. The movement flow is deliberate and purposeful. I place my hands the same way every time. I give a twist at the same instant every time. I'm feeling for the esophagus with my pointer finger every time. It's not some willy-nilly haphazard lackadaisical flailing around hoping that the viscera will come out without breaking the intestines. Successful farmers look at each task, at each day, with purpose. A successful farmer thinks while he does.

My brain is never disengaged. Even when I'm walking somewhere seemingly unoccupied, my mind is brimming with ideas and my eyes are searching for an errant piece of unsightly plastic to be picked up, a nail in the driveway that might puncture a tire, a garden tool misplaced from its proper rack. You've got to stay on task, stay focused, all day, every day. You can't afford to go into neutral.

Many years ago we would allow folks to come and tag along for a day or two. Always wonderful folks--and you know I love people--it was frustrating because it was distracting. In frustration, we decided one year to allow tag-alongs for a week if they paid us $1,000. Of course, everyone who wants to tag along believes they can help enough to earn their keep. "I can definitely pull my weight," they all promise. Did I say I love people? Yes, I do, and I don't want anyone reading this to think I don't want visitors or don't like people. I'm as extroverted as they come. People energize me.

Okay, so we instituted this new policy: $1,000 per week for tag-alongs. We had several takers that season but discontinued it at the end of the year. Even at $1,000 per person per week, it could not pay for the distraction the arrangement created. It actually made it worse because people paying $1,000 for a tag-along week felt more entitled to badger with questions. So my whole day, day after day, was consumed with answering questions and I forgot the pliers or forgot the garden hose fitting, whatever I needed to complete a task. At the end of the day I couldn't get done half what I was supposed to do because my mind was not on task. And I love people.

This is why we prohibit interns from using anything electronic when they're out working on the farm. They need to be 110 percent engaged. Distractions are not only wasteful, they can actually be deadly. We don't run tractors with cabs and radios in them. We don't listen to music while working. Use that time to think about how you could do the task better, or how to do the next task better. If you have time to listen to entertainment while working, you're not using your mind in real time to figure out how to do things better, or whether you ought to be doing what you're doing, or contemplating how to oust that idiot politician from power.

I heard an old saying that geniuses spend a lot of time walking. I think it's because something mentally energizing occurs when you're viscerally engaged in an activity. Your awareness is more acute than when you're just chilling out sitting in the hammock sipping lemonade. To be sure, I agree with sitting in the hammock sipping lemonade. And that might be a good time to listen to entertaining audio. But when you're actively engaged in work, it's time to think.

When putting out transplants, as monotonous as that may be, each plant should be placed just so. Whether you're a biodynamic operative placing the largest root north or whatever, placing them haphazardly, as if it's just a rote, mechanical process, will not yield the results that thoughtful, customized placement provides. Some farmers breathe a prayer on each one as it's put in the ground. Are you aware of the gold finch chirping--where is it? Can I catch a glimpse of it? If tasks can be done thoughtlessly, I'd say they're either not the right ones or not in the right context. I can't imagine a task that doesn't require mental engagement.

With all of this in mind, here are three practical ways to leverage our time and motion.

1. Set benchmarks. How do you know if you or someone you're training is efficient if you don't know the standards of efficiency? How long does it take to plant 100 onion sets? How long does it take to unload a hay wagon? Every task should have benchmarks of excellence.

To be sure, these benchmarks can change if you have an efficiency breakthrough. These changes will come over time and through repetition. But you have to know what they are so you can assign value to them. I'm going somewhere with this discussion, so hang with me. We're building on previous discussions, especially the one about assembling a team and creating fiefdoms with self-created compensation packages.

For fun, let's run through an example. If we're going to raise broilers, we have several phases: brooder, field shelter, processing. If we have a multi-income farm, a team-oriented outfit, we'll have infrastructure and interests that will want to do one of these or all of these. Maybe two of these. Maybe someone wants to brood chicks and not move shelters in the field. Maybe someone really likes the field work but doesn't want to brood the birds.

For example, we have one team member who lives in a rental house on a neighbor's place but caretakes one of our rental farms a couple of miles away, where no house exists. He doesn't have brooding capabilities where he lives so doesn't do that phase of the production. He takes the chicks once they're big enough to go on pasture and raises them on the farm he caretakes, which is about 4 miles from where he lives. Logistically, we need to figure out how to leverage the rental farm where no electricity and no house exists, while at the same time not complicating his living situation with the complications of a brooder. A discussion about brooders where he lives at the neighbor's place would complicate the arrangement. It's better to keep things simple at the neighbor's place where he lives. Because we know the value of brooding as a component of the whole production cycle, we can easily break that off in this particular arrangement since brooding is problematic where he lives.

Just like segmenting the accounting categories creates the opportunity for gross margin analysis, so segmenting tasks offers customized compensation packages for different situations among your team members. What if you have a partner who wants to raise chickens but not dress them? Or a partner who loves processing but not production? Maybe you have a friend that wants to be on the processing crew but can't do anything in the field. Having a fit, with designated value, for each scenario increases your chances of finding

Time and Motion Studies

a match and building a team, otherwise known as getting some help.

So let's assume a batch of 1,000 chicks. How much time does it take to care for them? Feeding and watering should take no more than 30 minutes per day--15 minutes in the morning and 15 minutes in the afternoon. If they're in the brooder for three weeks, that's 21 days at 30 minutes a day, or 630 minutes spread over 1,000 chicks. That's 10.5 hours and if our labor rate is $20 per hour, that's $210 divided by 1,000 chicks is 21 cents per chick.

Now let's go to the field. Our benchmark at Polyface is 60 seconds per shelter per move for 75 birds, but assume 70 saleable (mortality, sickness, predation). The feeding and watering benchmark is an additional 60 seconds apiece and if we have evening chores (the final 2.5 weeks) we have another 60 seconds for feed and water. Taking all this into account, a conservative benchmark is 5 minutes per shelter in caretaking time per day. They're in the field for 35 days X 5 minutes per day is 165 minutes per shelter divided by 70 birds per shelter is 2.35 minutes per bird. By the way, this is an extremely conservative benchmark. My personal benchmark is more like 3.5 minutes per shelter day rather than the above 5 minutes, so we have a lot of wiggle room here for novices.

Going with the 2.35 minutes per bird figure, that means we can handle 25 birds per person-hour in the field. Returning to our $20 per hour base labor rate, that's a direct cost of 80 cents per bird. So if we pay a contractor (team member, fiefdom operator) $1 per bird for the pasture production segment, that's a nice, generous labor rate. If that person also wants to brood, we can pay $1.21 (add the brooding rate) per process-ready bird. If we want to pad it a little, call it $1.30 or even $1.50. Be generous and you'll get great help.

Now what about processing? Our benchmark is 20 birds per person-hour. At $20 per hour labor base, that would be $1 per bird. Now you can add these up for someone who wants to do the whole shebang, start to finish, or you can have three different people doing the three different stages. The ultimate mix and match. More importantly, it offers lots of choice among our team members.

Returning to the gross margin analysis and pricing, now you can create a defensible and precise total expense understanding so you can price knowing that you're in a positive margin. Obviously all sorts of things can affect these benchmarks. Automatic nipple waterers, for example, can cut brooder chores in half. Increasing the number of birds in the brooder lowers the per labor cost.

Why? Because transporting yourself to the brooder is the same whether you're going to service 500 birds or 2,000. Opening the door, stepping inside, checking things are all the same regardless of how many chicks you're servicing. Economies of scale is a real principle, even on small farms. Remember, the cost is in being there. Doing more while you're there adds little time to the chore.

Handiness helps everything. If it's on a remote farm where you have to drive, that adds time. If it's outside the yard fence, that drops time. Clumping more things together saves time. Spreading things out requires more steps or more running, what we call the cost of being there.

A good example of this is our eggmobile. It's a portable hen house that follows the cows. The chickens free range out from it, scratch through the cow dung, eat out fly larvae (maggots) and spread the dung over more ground, acting as a fertilizer spreader. My progression was a systematic economies-of-scale trajectory that provides a good example of the kinds of compromises we make between altruism and business sense.

My first eggmobile was only 6 ft. X 8 ft. on bicycle wheels with a hexagonal moveable yard built out of 12 ft. chicken wire panels 5 ft. tall. I had two pairs of gate triplicates wired together. The yard rotated around the eggmobile like leaves on a four-leaf clover and remained upright due to the doglegs in the hexagonal configuration. When I finished going all the way around the eggmobile, I'd shove it up the field and make a new four-leaf clover.

One day I happened to intersect with where the cows had been and my epiphanal lights went on--because I was observing and thinking and not distracted by ear buds or headphones, by the way.

The chickens attacked the three-day-old cow patties. Aha! Chickens following cows.

I retrofitted the eggmobile to a 3-point hitch so I could move it longer distances. The bicycle deal was okay, but I couldn't shove it all over the farm. I let the chickens free range and they obliterated the cow dung. I knew I was onto something. But the 3-point hitch arrangement was cumbersome. The next step was a trailer eggmobile. I built one 12 ft. X 20 ft. long and put in 100 birds. It worked beautifully and the chickens literally lived off the land. I didn't feed them anything but some corn for energy. The eggs were amazing.

People wanted more eggs. This is long before electronic door closers had been invented. The eggmobile required me to go up and move it with a tractor, collect eggs in the afternoon, carry the eggs home (sometimes from nearly half a mile away) and then go back after dark and close the doors so foxes and raccoons wouldn't get in. Meanwhile, customers wanted more eggs.

Even though the birds were eating almost nothing and performing tremendous sanitation and ecological benefits to the pasture, I still had an hour of time in these eggs. Moving, closing up, toting eggs home--five dozen eggs get heavy when you carry them for half a mile, especially after making hay in the afternoon. At four or five dozen, that hour was a lot of time. I needed about $25 a dozen to make it work.

Nobody would pay that--at least not very many. So I elected to add more birds. With 200 hens, my labor efficiency went up, but so did feed consumption. Now the birds needed feed because they didn't range any farther than the 100 and there wasn't enough gleaning over that ground to satisfy 200. So now I had to supplement with feed.

Today, we run two eggmobiles hooked together with 800 birds. Yes, they eat a lot of feed, but now we can afford to run an ATV to get the 40 dozen eggs. We still get all the benefits but now the labor cost-per dozen has dropped ten-fold. You could argue that the eggs are not quite as good as they were when the chickens were just living off the land, and I would agree. But they are immeasurably better than industrial alternatives and now I can price them competitively while still maintaining a good gross margin.

I remember well when people castigated Eliot Coleman for using propane in his greenhouses. He said saving the planet had to take a back seat to staying in business and feeding his community. I would argue that the propane offsets the diesel fuel to bring greens from warm climes to Maine in January. So at worst it's a wash; if you studied it, I'd say some propane is a lot less than the transportation energy, especially when you add in the cost of road maintenance. And I think even Eliot would agree that in the perfect world, every home would have a solarium attached to the south side so each family produced their own greens in Maine in January--with enough passive heat left over to supplement their house comfort.

I've made it clear that in my perfect world, we wouldn't even sell eggs. As Pat Foreman, ultimate chicken whisperer has pointed out in *CITY CHICKS*, if roughly every third household in America had a few chickens to eat the table scraps from the three households, that would produce all the eggs the entire country needs. Who needs egg factories? She's absolutely right. If all the people with big hearts who demonize factory eggs would push this in-house integrated agenda instead of pushing a vapid factory farmed cage-free option, we'd get where we should go much faster. But I digress.

Once you have your benchmarks, you can create your compensation platform that then becomes the foundation for your memorandums of understanding (MOUs). The beauty of these benchmarks is that they can establish a value without the tension of hourly wages. As I've gone through this analysis, can you imagine throwing it all out and just hiring someone for $15 an hour? Goodness, all you'd do is sit there shaking your head at the inefficiency. Where's the drive, where's the efficiency?

To be sure, sometimes people come back and say the compensation isn't enough. Sometimes we've missed something, but more often than not, it's just because they don't want to work that hard.

Time and Motion Studies

My answer is why should I pay you more for doing a job just because you're slower? Either get with it or do something else. We always set conservative benchmarks so if someone exceeds them, they can make really good money. An exceptionally efficient pastured poultry operator, for example, using the benchmarks I've established above, can earn $30 an hour. That's not bad pay for farming field work.

I hope by now you can see clearly why hiring people by the hour is a dead end. If you have a person who just wants to work by the hour, chances are that's a person you don't want. Why in the world put yourself through the angst of watching inefficiency on your dime? On the other hand, why deny the go-getter the chance to progress to a much higher hourly return?

I hope this all makes the relationship and synergy of additional people, margins, and efficiencies come together. It's a different way to look at help; a different way to structure the business, but it serves as a protection as well as a promoter.

If you're running a produce operation, put your help on performance quota. This is how orchardists have paid apple pickers forever. So much a bushel. Can't we adapt that to tomatoes? I'm convinced that we smaller farmers need to be more creative at duplicating some of these piece-work ideas developed by the big guys. The reason they work is because it saves the bosses from the consternation of seeing one worker doing double the work of another but getting the same hourly wage. That's not fair to the go-getter. And it protects workers from time-creep, the temptation of everyone who works for wages. "Just putting in my time" is one of the most common phrases in the work force. It makes people who have invested their whole lives into a business, worked 100 hours a week for pennies to get it launched, and continue to scrimp personally for the good of the business go crazy over frustration with the help problem.

Give people responsibility over their own sector, compensation that rewards their personal competence, and get out of the way, or at least enjoy just watching from afar. I'm reminded of the time and motion studies Andre Voisin, the great guru of grass farming, performed in France. He sent students out to the field to follow cows around, one student per cow, and count the number of times each cows' jaw moved. He wanted to know how many times a cow chewed in a day. To his astonishment, he found as much as a 50 percent difference, within the same herd and same field, in chews per cow per day.

Not surprisingly, the cows who chewed more put more milk in the bucket. Rather than just being some genetic metabolic genie that created high producers and low produces without regard to measurables, he found that the higher producers actually worked harder. How about that? Every farmer knows that we have doer animals and doer plants. We also have poor performing animals and poor performing plants. Much of farming success depends on aggressive culling to concentrate our efforts and resources on productive plants and animals, not the weak and sickly.

You see where this is going. Folks, you need performers on your farm. A person with a negative attitude or a lazy bent will not change regardless of incentive, brow beating, or whatever. People ask me how we pick our interns, and I always respond with a single word, "Attitude." I can teach anyone how to gut a chicken, but I can't teach him how to like it. That's attitude, or spirit. I'll take total ignorance with good spirit any day over an educated bum. And if you want a successful farm team, you'll do well to empower the one and get rid of the other. Any business person, including me, will tell you that some of our biggest mistakes over the years was putting up with poor performers for too long.

The beauty of benchmarks with time and motion studies is that you don't have to go through the angst of these difficult discussions. What we've found is that the efficient folks thrive and the inefficient ones leave because they can't reach the benchmark. That's liberating. As my good friend and entrepreneur coach Lyons Brown says, "go-getters are drawn to accountability. The rest are scared off by it."

Carry a stopwatch, do the math, write down the benchmarks. Set value. Then if you need another team member, you can offer something concrete and precise. If you have a soul who wanders in and wants to help, you have the same offer toward efficiency. It's ultimately not a confining thing, but a freeing thing. So the first practical time and motion advice is to set benchmarks. Now to the next one.

2. Go loaded and come loaded. This is such an obvious point that it seems almost silly to mention it, but I'm constantly surprised at how little it's followed on the average farm. Much of farming is toting stuff, either physically or with a machine.

To reiterate, the big cost is in being there. It's the movement from point A to point B that creates the expense. That's time and fuel. Certainly the trucking industry understands this. Do you know how cheap you can get something delivered if it's on a back haul? With sophisticated logistics software now, fewer and fewer trucks are getting stranded with nothing to haul back home, which is bringing overall trucking costs down.

It's amazing how cheap transportation has become. Fuel isn't cheaper. Trucks aren't cheaper. Labor isn't cheaper. What has enabled the cost of trucking to actually drop over time is tight logistics that reduce empty truck movement. Remember how many empty trucks you used to see on the interstate? You don't see them today. Every truck is loaded. That means we're hauling more freight with fewer trucks and the cost per pound has dropped.

On the farm, we're often in the moving business. Whether it's toting carrots from the garden, hay from the field, or logs from the forest, we're often in the materials handling business. We need to think like a trucking company. What can we carry both directions? It might not be something we need today, but look out a week and think. Surely times exist when we have to dead head one direction; we can't carry stuff all the time. But an extra trip is an extra trip.

This includes trips to town. One of the quickest ways to cut into your profit is too many trips to town. Wait until you've accumulated several errands before heading into town. If you do need to go to town, what else can you hang around the trip? This means we have to own vehicles that are multi-use. I'm a fan of little, maneuverable trailers that we can tow behind the car or SUV to haul stuff home. Go to town and unhitch the trailer at the farm store. Do your errands, then go back and hitch up, get your stuff, and head home. In the big scheme of things, the inconvenience of towing a light trailer and unhooking it is more than offset by eliminating another trip to town with a designated hauling vehicle.

The big savings here is not the fuel as much as it's your time. Often farmers who complain about never having time to relax fritter away most of our time by being inefficient with our running around. Always think about what you can add on to a trip--anywhere--that will make it do more with less.

We drill into our interns to never walk with one bucket. First, it's off-balancing, but second, if you're going anyway, take two. If you don't need two today, leave one so the next day you don't have to carry anything. It's always more efficient to take a tool box full of tools than to take just the ones you think you'll need, only to find out at the jobsite that you need one more tool.

We have a couple dozen big plastic totes. They're great for organizing your stuff and provide enough room to take plenty instead of just enough. You can always bring a half-full tote home much easier than running out of staples two posts from the end of a fence. When we go move cows, we take a tote with a couple of insulators, a hose valve, a little piece of splice wire, pliers--you get the picture. Assume something is wrong and chances are, nothing will be. The very time you get cocky and say to yourself, "Nothing's been wrong for a month. I'll just run over there and move the cows without all this stuff," guess what? That'll be the very time you need the tote.

A corollary to this principle is to handle stuff only once. This includes emails. I struggle with this one. When you open an email, deal with it right then--answer, delete, or file. I learned this from a friend, long before email was around, who said he always tried to handle mail only one time. When he opened a letter, he found that if he answered it, threw it away, or filed it somewhere immediately, it enabled him to keep up with stuff and keep his work space clean. Filing away does not mean stuffing the letter back in the envelope and throwing it on the corner of the desk. I admit to extreme guilt on this one. But over the years when I do catch up and adhere rigorously to this concept, my life gets easier. It's true.

But then a string of perfect 70 degree days come along, lots of outside work to do, a bunch of mail, and suddenly it's more than I can handle in an evening, and then . . . You know the drill. Aaaaargh! But that doesn't mean I can't aspire.

Do like things together. Gather all the eggs quickly. Don't worry about separating cleans from dirties or sizing them when gathering. I know this may sound contradictory to the handle only once concept, but product flow often means a kind of assembly line mentality. We do handle stuff a second time, but you don't want to do the same procedure a second time. When we're gathering eggs, we don't look at them: gather as fast as you can. Then when you get to your processing station, you have scales, damp cloth for wiping little specks, and you're only actually looking at them once.

When we accumulate our product for orders, we create a pick sheet (40 packages of T-bone, 100 whole broilers, 20 pork tenderloin, etc.) and back up to the walk-in freezer with the box truck one time. We load everything from the pick sheet and bring it into the work room at one time. We use plastic totes (like milk carton totes). One has T-bone steaks, another chicken livers, another pork tenderloin filets--you get the picture. It's all assembled on tables so we don't have to stoop over (ergonomics is all part of time and motion) as we collate the orders. We don't take coolers back to the walk-in until everything is filled. If we took them to the walk-in as they were filled, we'd lose precious order assembly time. This is a perfect example of clumping.

One run and stop to the walk-in, one run and stop to the work room, one session in the work room, one load up back onto the empty truck still parked at the work room, one run and stop back out to the walk-in, where the truck waits overnight to be loaded in the morning for the delivery run. What eats your time are lots of individual runs back and forth to the freezer. Locate your critical infrastructure like this so it's close. Spreading out infrastructure can create a lot of lost

241

motion in a year. Concentrate to the hub.

Except. When Daniel and Sheri got married and built their house, we started thinking about where it should go and realized that our main farmstead hub is not centrally located on the farm. The original farm house is out on a corner end. We called the power company and asked how far they would run power to a new customer without charging anything. They gave us a ballpark and we picked a location as far as we could go without paying for the power line.

That moved hot power several hundred yards farther into the farm than we'd ever had. A couple of years ago when we finally had enough winter snow-melt water impounded to justify investing in a New Zealand K-line irrigation module, we were able to fairly cheaply extend a double-ought power line from Daniel and Sheri's house out to a pond to cheaply run a 20-volt irrigation pump. That would have been a prohibitively costly project beforehand, but having that hot power extended farther into the farm made it affordable.

Furthermore, since their house was there, we ran water from our well, which is very strong, to their house. When we needed more hay storage space and winter feeding space, that nice well water offered a great option for locating the new structure toward the far end of the farm rather than here near the house where the old hay shed/feeding complex is located. Essentially we've used Daniel and Sheri's house as the basis of a new farmstead nucleus on the other end of the farm.

While it does create some duplication, it is far more efficient. Prior to this, hay made half a mile away in those far fields had to come all the way to the original farmstead end. The compost generated by feeding that hay had to be toted all the way back to those far away fields. Now, the hay goes into its nearby shed and the compost stays close too. This is a little bit like Ben Hartman's idea of duplicating little tool sheds around his produce operation.

On this note, let me tell you one of our most satisfying stories. When GMOs came out, the local feed mill we were using scoffed at our desire to have non-GMO feed. We looked around and found a little defunct feed mill about 15 miles away, located on a farm owned by a Mennonite family. I asked them if they'd mill feed for us and the 40-ish year old son jumped at it. It turned out that the old late 1950s-style mill had become a source of tension between father (who installed the mill to grind his own chicken feed) and son.

When the poultry industry consolidated and vertically integrated, the mill became obsolete because the poultry integrator brought its own feed to the farm. The mothballed mill had not been used for years, but the dad wanted to keep the building painted and nice. The son wanted to tear it down. They started milling our feed and then other people started coming to them and our volume picked up. We ran a couple of four-ton Grain-O-Vator (great company, by the way) feed buggies at the time. We towed them behind the pickup and made the 30-mile round trip once a week, then twice a week, then three times a week. After a couple of near-fatal situations (four tons of feed on a pickup bumper hitch can be exciting), we started looking at options. The final hair-raising incident involved a school bus stopped at the bottom of a hill letting off kids. That was not a good situation to be too heavy to stop with the loaded feed buggy. Time for changes.

I asked the family if they had ever considered a bulk truck. Oh, they responded, they dreamed of one every day, but were too leveraged with the fledgling and burgeoning (growing businesses are always strapped for cash) business to afford a bulk truck. "What if I gave you a three-year loan, at no interest, to buy the truck? You can pay me back monthly, annually, or wait until the last day of three years. I don't care. What do you think?"

They thought they'd just gotten manna from heaven. Within a week they had found a great truck in North Carolina. I wrote them a check, they picked it up and suddenly we saved two whole person-days a week, not to mention wear and tear and desperate prayers for safety. It was by far and away the best investment we ever made. That little company, Sunrise Farms, now is the go-to non-GMO feed supplier in the entire mid-Atlantic region. That truck enabled them to step up and service larger and farther accounts. Many businesses saw wins with that decision, which was really nothing more than a time and motion decision.

Obviously the idea of go loaded and come loaded has lots of ramifications for efficiency, and it's all facilitated with the final practical tip:

3. Make lists. I don't know about you, but I'm a list maker. My wife Teresa is an even better list maker. Organizing your day, your week, your month into a written, prioritized list is probably one of the most efficient things you can do. Juggling all the chores and projects necessary in a day can leave us frazzled, knowing we forgot something.

But a clear outline of what we have to do creates order out of chaos. It's also one of the best ways to go loaded and come loaded. It helps us visualize the day so we can organize it into a cogent flow. Because thoughts can be fleeting, it's good to be close to pencil and paper at all times. I guess smart phones fill this space today. At any rate, how many times have you had a thought and then at the end of the day try as you might, you can't remember what that good idea was ? It's maddening, isn't it? Write it down. I use scrap envelopes. They're kind of heavy duty and I can fold one up in my pocket with a stub of pencil and it works great.

I don't care if you use little spiral pocket pads, smartphones, or whatever. But keep a way handy to write things down. At Polyface, we make a long-range list--like a year or two out, then a monthly list, then a weekly list and finally daily lists, plus one more, and perhaps the most important of all: the filler list.

A filler list is stuff that requires less than an hour to complete. You know how it is. You get to a point in the day and look at your watch and think, "Not enough time to start on something and too much time to waste." By the time you work through that tension, another ten minutes have ticked away and you say, "Oh, forget it. I'm heading on to the house."

If you add up all the little snippets of time like this, in a year, you'd be surprised how many days you'll find. This filler list is fluid, of course, and changes weekly. Just having all these little jobs written down ensures that all those little edges of time get used productively.

This list making is not lost time. It's WOTB (Working On The Business) work, which has the most impact on your business. The list needs to be circulated to everyone involved in the business because it's the action steps--the bites, if you will, defining how we're going to eat the elephant. It stimulates us emotionally to see things broken down. If you have a huge project, don't just put it on the list. Break it down into many segments: prepare site, put in poles, mill lumber, put on roof. It makes for a longer list, but it makes the increments easier to visualize.

Once you have the list, prioritize it. A non-prioritized list does not create direction. It's a smattering, which is helpful to see, but not the way to focus. The list translates into focus and movement once we prioritize it. Your priorities will be based on different things. Sometimes littlest to biggest is a good way, creating a sense of movement, which is always energizing.

Most commonly, the priorities are about most critical to least critical. We have projects we've carried over from one year's to-do list to another for several years. These are want-to projects, but not have-to. Organizing it like this gives focus.

And then there's time versus money. If we're longer on time, we tend to do the time-heavy projects first. If we have some extra money, we might knock out some of those. You can use any criteria you want to organize the list, but if you want it to be useful, go the little extra step and prioritize it. Teresa and I have both been known to get to the end of the day and realize we've done a couple of critical things that came up that weren't on the list. We'll write them on the list so we can scratch them off. Few things offer as much satisfaction as scratching through a to-do item. Can you do that on a smartphone? I don't have a clue. I'm afraid I'm addicted to paper and ink for the rest of my life.

May I share one more story? Just yesterday Daniel told me that an industrial wood chipper we'd rented for a couple of years had been listed for sale. Few things fire up my imagination and desire more than something that makes biomass easier to acquire. We ran low on barn bedding last week and had to call for a tractor trailer load of bark mulch, the cheapest biomass we've been able to find: $1,200. That load is 90 cubic yards. The going rate of bedding-acceptable biomass, therefore, is $13 per cubic yard ($1,200 divided by 90).

That gave us a benchmark, or a market price. That's the default price when all else fails. We already owned, and have for a long time, a Valby 3-point hitch chipper that can handle 9 inch material. With that machine, we routinely chip 2 cubic yards per man-hour. If that's worth $26 ($13 X 2 yards), suddenly our labor is worth well over $20 per hour, figuring wear and tear on the machine, and fuel. But when we chip our own material, we get multiple benefits: upgraded woodlot, clean area, pretty woods, etc.

With this bigger machine, since we've rented it for a couple of weeks the last two years, we know we can generate 9 cubic yards per man-hour. At the market rate of $13 per yard, suddenly our labor value is $117 ($13 X 9 yards). Take off a little for depreciation and fuel, and you're easily at $100 per hour. This analysis showed quickly that chipping for ourselves rather than buying from a commercial source was time well spent. Looked at another way, purchasing the biomass was $13 per cubic yard; using our 3-pt. hitch Valby the cost was $10 per cubic yard; with the new super duper Vermeer machine, $2 per cubic yard. If you're using 10,000 cubic yards a year, the efficiency of that big machine suddenly looks amazing.

Since this is our fertilizer budget, it's also money better spent. The question is whether or not we use enough biomass to justify the bigger and more expensive machine. Suddenly, the decision to buy the big machine was a no-brainer. We bought the machine and we're in love. If we just stodgily went along buying loads of biomass because we didn't want to buy machinery as a matter of principle, we'd be shooting ourselves in the proverbial financial foot.

Before we did this analysis, I actually thought buying biomass by the tractor trailer load from a nearby sawmill was the cheapest alternative. It was easy. You pick up the phone, write the check. But convenience often comes with a price. By sitting down and crunching the numbers, doing the timing, then doing the analysis, you'd be surprised how many times prejudices or intuitions are actually wrong. So do the analysis.

There you have it, folks, time and motion studies. I hope you realize now that when we talk about efficiency, we're not talking about intuition. We're talking about hard core records and analysis.

We're talking about never being satisfied, but always looking for that one additional refinement that will enable us to do something better, easier, faster. As the old saying goes, working smarter, not harder. Don't be fooled; it's still plenty of hard work. But hard work is much more satisfying if we know it is the right work, and done as efficiently as possible.

Chapter 13

Getting Started (or Starting Over)

Everyone comes to this farming vocation under different circumstances: age, finances, talents, character. A clear assessment of where you are and what you can bring to the table is critical in order to create a way into farming.

In my experience, I see three groups of people. Each comes with assets and liabilities. Let's examine each group, and I'll talk to each like I'm sitting across the table having a conversation with you. Imagine you've come to me with this idea to start farming and you've asked my counsel about how to proceed.

MID-LIFERS

The first group is the mid-life crisis successful professionals age 40-55, with late teenage and young adult children. Generally urbanites, you folks have drunk the Kool-Aid from folks like me describing the idyllic life of pastoralism and family-friendly business.

You look at your rat race, the vapid legacy of your career, and yearn for something more meaningful, permanent, and satisfying. Often you have a strong desire for self-employment, to be your own boss for a change, and to escape the dramatic interpersonal soap operas of office politics and corporate schmoozing. It's as much a plan of escape as a plan for progress.

Getting Started (or Starting Over)

Generally you portray success in your career and finances. You've probably paid off your mortgage, accumulated some retirement and a healthy nest egg, and climbed the corporate ladder into mid-management at least. Except for the frustration of the frenetic and boot-licking office, you live a relatively comfortable life. You can order take-out any time. Paid vacation, paid medical, sick leave--the package is pretty cushy.

You've conveniently forgotten your early days of putting in longer hours with fewer benefits and perks. Except for the vexation of spirit at your workplace, by almost any measure you've got it made. In fact, it is precisely this comfort, this new ability to daydream, that affords you the capacity to indulge in a farm fantasy.

Tending calves and growing carrots sounds therapeutic. With nothing more to go on than a farm in the mind, you imagine healthy plants and animals, bucolic landscapes, puffy white clouds against a vibrant azure sky, a slight breeze and 70 degree days. Garnishing this picture with family, friends, grandchildren--oh, it's paradise on earth.

Now that we've built ourselves up into a frenzy of perfection, let me pop your bubble. Every day is not 70 degrees. Every day is not healthy. If you have animals, they need to be looked at EVERY day: Christmas, Easter, Sunday, your birthday. Not some days, EVERY day. When you're sick. When you'd rather be at a wedding or on vacation. EVERY day. Have I made that clear?

And herein is the pitfall for the mid-life group. Without exception, you underestimate the effort required for success. The ten years and 10,000 hours you put into becoming skilled at your vocation WILL NOT TRANSFER TO A NEW ONE. I put that in capital letters because it is by far and away the largest misconception of this group. You assume that the skills that enabled you to succeed in your current vocation will automatically make you skillful in farming.

Ain't gonna happen. When he was alive, Allan Nation and I used to share stories about how astute business people act like financial kindergarteners when it comes to their farm businesses.

Any fool can see they're throwing money away on their farm, but they can't see it. This is common enough to be practically axiomatic.

It's almost like successful first-career folks are brain damaged when it comes to a second career in farming. The learning curve does not change; it is true for every single vocation. Thinking you can short-cut the experience toward mastery is the universal downfall of this age group. If you fit this description, then, heed my warnings and advice here. Don't dismiss it as applying to all those other people, assuming you'll be different. You won't be. I've watched hundreds if not thousands of wanna-bes in this mid-life period jump ship, get frustrated, then blame the farm for their frustrations and financial losses.

As a side note, realize that if you don't get your children on the farm before they're 10 years old, you will have a hard time getting them on board. Asphalt, hanging out with friends over take-out, and the urban social scene mold the psyche of the young extremely early. Again, I've watched hundreds of families yank their teenagers out of the urban environment to go live Mom or Dad's dream on a farm. Rather than embracing this new situation, the children resent it. Farms are not reform schools. If you're having tension with your teens in the city, chances are the farm will only exaggerate them, especially if the parents are tense because the farm isn't working as smoothly as planned.

With all that said, this mid-life professional group has some strong countervailing assets to bring to their farm table. The first is life experience. By now, you know some people are crooks and others are honest. You know every proposition is not as good as it sounds. You know it takes time to develop relationships and to build a team. You know people are different, with all sorts of complementary and competitive skills. Life experience is valuable and you can be thankful you've acquired it.

Second, you have connections. You know people. People at work, at civic organizations, church, or charities where you've volunteered. All of these connections can be leveraged to find help, to develop customers, and to find deals. You might know the folks down at the planning commission or zoning office who will need to sign off on a farm project. Knowing people is a huge asset when you're launching an endeavor.

Third, you have capital. I spend a lot of time with 18-24 year-olds who want to farm--they'll be another group--and their poverty is a universal vexation. Saddled with student loans, trying to get started, their lack of capital is a real challenge. But you mid-lifers have some capital. You might even be able to buy 100 acres without a mortgage. When I tell you about needing $20,000 to install some fence and water line, you consider it chump change. Goodness, you're earning $80,000 a year, so you're used to generating $20,000 in three months. No problem, you grin.

And therein lies the danger. It sounds too easy. Because capital is readily available, you're not poor enough to drive fiscal creativity to succeed before the financial hemorrhage bleeds off your capital. Anyone who isn't scared to death of running out of money isn't ready to launch a farm. A $30,000 tractor here, a $15,000 baler there, a $10,000 attachment to lay plastic--your pocket can be picked in a minute.

The commonality here is that often none of these expenses is necessary. The 22-year-old wouldn't think of buying this stuff; they borrow, rent, or buy custom work because they can't afford to spend that much money. You, on the other hand, with your fancy bank account and nest egg, buy all of this stuff before you even have enough experience to know if the purchase is necessary. I've watched people buy a chainsaw before they even knew how to sharpen it or use it. Ditto for tractors, ATVs (All Terrain Vehicles), and a host of other things.

Allan Nation used to always say that profitable farms have a threadbare look. Board fences and manicured fields indicate what Allan used to call a "land yacht." A farm can be a money pit much easier than it can generate income. Having capital to start is a wonderful blessing on the one hand, but a definite curse on the other. Wasting money is a big problem in this age group.

The other big challenge is physical energy. I think I read somewhere that after 30 years of age we lose a couple percent of our cardio-vascular capacity every year as part of the aging process. You won't live forever. Physical acumen wanes; it's a fact of life. You won't be able to put in the work day at 40 or 50 that you could at 20. As much as we'd like to think we'll be immune from this, we won't be. The later in life you come to the farming vocation, the harder it will be to put in the long days necessary to launch the business.

In fact, I advise people older than 55 to NOT start farming. Instead, find a young person on whom to leverage your connections, capital, and business savvy and live this new fantasy vicariously through the energy of a youthful partner. Plenty of 55-year-olds have taken me to task for this, but I stand by it. By the time you have your 10 years and 10,000 hours, you'll be 65. You think you get tired at 55, let me tell you; 65 will be far worse.

So what's a mid-lifer to do? Here is my recipe for getting in. First, protect your cash. Do something where you are--become self-reliant on your current footprint. If you aren't growing vines on your house, garden in your lawn, bees on your roof, and chickens in your house, forget about farming. What you do now on this trajectory is the way for you to determine if it really is your passion or a passing fancy. Is your farm fantasy just a reaction to a bad day--or bad year--at work?

If your teenagers won't gather eggs where you live now, they won't gather eggs on a farm. I call this filling up where you are. Plenty of production can and does occur in extremely small spaces. Do it. If you can't wait to get up in the morning and tend the veggies growing in tubes on your front porch, you won't mind jumping out of bed to tend an acre of veggies on your farm. By the same token, if you quickly tire of caring for your urban or suburban plant/animal pursuits, you'll tire even quicker out on the farm. Get it?

Like good marriages, practice and proximity actually increase love and affection. If your on-current-location production increases your yearning for more, then you're probably cut out to be a farmer. But if after a year you find your eagerness to tend the plants, chickens, and earthworm bed beginning to wane, you've just made a wonderful and financially rewarding discovery about yourself--you're not cut out to be a farmer. Better to learn this now before going through major losses of money, time, and relationships, don't you think?

If after a year or so of filling up your space you're even more gung-ho to proceed, then begin looking for opportunities. I won't belabor all the leasing, buying, and layering options here because I've done that in other books. For now, I want you to focus on practicing the trade as much as you can where you are with what you have, right now. If you find yourself not watching ESPN any more because you're tending your hanging patio garden, you just might have what it takes to be a successful farmer. On the other hand, if you have weeds in your barrel garden or your backyard garden, you'll have way more on your farm.

If you dread cleaning out that chicken coop in your urban setting, you'll dread it even more on the farm when you have 1,000 chickens instead of 10. Your most important task is to test your farm dream before you change your context. Everything you can do to determine if this change is right, without risking your relationships and finances, will be valuable. Your life has been different than farming. Your career path has not been in farming; why change? Is this hormonal? Is it a phase? Is it a reaction? Your most significant discovery needs to be about you and to find out if in the deepest part of your soul you really were a farmer and got sidetracked by life circumstances. If returning to the true you means to start farming, wonderful. If not, that's okay; not everyone can or should be a farmer.

How-to information is everywhere. Cows, sheep, strawberries, peaches--we've never had as much access to how-to information. I don't think that's lacking. What's lacking is the personal assessment to know if you're cut out to be a farmer, and I hope this discussion can help sort that out.

YOUNG PROFESSIONALS

You are 28-40-year-olds and already burned out in your Dilbert Cubicles. In the last decade, you have come to dominate our Polyface intern applicant program. It's a burgeoning pool that is making up a larger portion of farmer wanna-bes.

I'm always impressed at *Mother Earth News* fairs with how young the audience is. And I don't think it's because I'm getting older. Also known as Gen-Xers and the front edge of the Millenials, you young people grew up with computers. Most have never even seen a typewriter.

Career-oriented, saddled with college debt, and often single but looking, you have neither the life experience nor the capital of the mid-life group. But you're younger, enthusiastic, and usually less tied down. This group could be subdivided into the ones with family and the ones without family.

Marriage and children, which normally occur during this age, fundamentally alter your circumstances. In the analysis of stressful life events, only a few things are 100 percent: marriage, change of jobs, change of address, death of immediate family member, divorce, birth or adoption of a child. The rule of thumb is that you can handle one or two of these in a year, but three is too many.

In our intern and apprentice program here at Polyface, we've certainly corroborated this notion. With so many people living together prior to marriage, that life event may wane over time as a major life-altering stress situation. But since our farm tends to cater to folks of a more conservative background, in our experience newlyweds and internship don't mix well. It's too much new all at once. Divided interests are problematic for both the marriage and the learning.

Add children to the mix and we have interns running off to skype their kids instead of being part of the team's after-supper hang-out conversation. These same pressures are there in a start-up farming situation. I've described what we see in our internship program, but if you're launching a farm business the stress factor is even higher. This is why I'm adamant that you be debt-free and have one year's worth of savings prior to leaving your off-farm job. I'm completely in favor of farming full-time by leaving your off-farm job, but timing is everything.

We have to be careful about the demands life puts on the pieces of our life energy. Your spouse wants a piece of you. Your children want a piece of you. Your business wants a piece of you. If financial and emotional stress keep you from being able to provide enough of these pieces, life begins to crumble. Your inner compass begins to waver. Don't do that to yourself. Appreciate proper timing and where you are in the life cycle.

Children go through a life cycle too. An infant is far different than a 5-year-old. No one is a bigger fan of farms as good places to raise kids than I am. My book *FAMILY FRIENDLY FARMING* details principles in this regard, including how to maintain family business harmony. I think coddling children is probably more abusive than putting them to age-appropriate work. Far more children have become dysfunctional adults through coddling than chores.

Think about where your children are in the cycle of life. Consider delaying the full-time farm leap for two years to let that toddler get potty trained and talking. It will make the whole transition easier. Life will proceed fast enough without throwing additional stresses into your playing field. One or two years, more or less, won't make a big difference in the trajectory of your farm business, but it could make a big difference in the journey of your marriage and family health.

Many times the best thing you can do is wait a little. Patience is a virtue for a reason. Spend your time preparing. Grow some things--anything--where you currently live. Attend seminars put on by people you want to emulate. Read, always. Visit farms. Participate in your tribe; develop connections; run ideas by experienced people--and then listen to their response.

If you're in this age bracket without children, you're somewhat freer to do the experimenting, traveling, and sleuthing I've suggested. Get your bills paid off; quit partying; start saving. Live cheaply; quit eating out; sprout your own mung beans. The problem with singles in this age bracket is that you are used to your freedom and lack of attachment to live as if everything is carefree and always will be. Realize that if you find it hard to save now, you'll find it even harder when you're farming and you aren't receiving that weekly paycheck.

The habits you create now will be the habits you carry over into your farming venture. You can't spend like a drunken sailor now and expect to suddenly become frugal when you begin farming. Cultivate early rising and early bedtimes. Throw out the video games and spend that time tending your plants, even if they're under grow lights in your apartment. Make sure they're edible. ha! If you're going to succeed, you need to be borderline obsessive compulsive. You might have to change friends. Get used to it; farming is not known for its social prowess.

This is the age when most new farmers launch. In my humble opinion, though, the best age is a decade younger.

YOUTH AND 20-SOMETHINGS

I came back to the farm full-time September 24, 1982 when I was 25. Except for college, I never physically left home. In the classic book *THE MILLIONAIRE NEXT DOOR*, one of the key ingredients to acquiring wealth is starting early--very early. The most physically aggressive time of life is your 20s. At no other time in life will you have more energy.

People who wait until their 30s and 40s to launch their farm business can never overcome that lost decade of physical prowess. Youth can, as they say, "take a lickin' and keep on tickin'." During the height of the European guild system, apprentices began around 10 or 11 years old and became journeymen during their late teens or very early 20s. This is the trade trajectory that is historically normal. People who are 30 year old and wondering about what they want to do when they grow up are historically abnormal.

I believe strongly that our modern American culture which denies, either socially or legally, most youthful interaction with meaningful adult work and relationships has created this vacillating young adult conundrum. The more different kinds of meaningful (notice I'm using meaningful) things a child or teen does, the more life experience accumulates. That's a good thing, not a bad thing.

Hanging out, partying, and goofing off have their place, but had better be limited if you really want to go places.

In our Polyface intern program, we work with a lot of you in this age category, of course. It follows like day and night that the ones who find little jobs to do after supper, for example, are the ones who excel at catching onto things during daylight work hours. The ones who jump in the car to run to town for an ice cream and a movie are the ones proverbially asking stupid questions and not "getting it" during the work day. Why? Because where the heart is determines where the mind is.

A word about college. I'm not opposed to it. But I think life experience is just as important. What would or could you do if you took those four years and used it in another way? What if you lived in another culture for a year? What if you did several apprenticeships? What if you launched your own business? What if you studied by correspondence? The notion that bright kids graduate from high school and then go to college is ingrained deeply into our culture. Unfortunately, that path is often a waste of time, waste of money, or both.

Today's educational choices are vastly different than a few decades ago. The internet has opened up distance learning like never before. Why not start a literary club in your community? Why does the intellectual stimulation offered by academic settings have to occur only in institutions charging you $30,000 a year to sit on their sofa? The Uberization of education is progressing fast, and I'm frankly glad to see it. Every community has well-spoken, well-written folks. Start something.

When I read the *LITTLE HOUSE ON THE PRAIRIE* books documenting community spelling bees, poetry reading, original essay contests and public debates my heart yearns for that kind of intellectual/academic stimulation today. It's been replaced with video games, TV sit-coms, and the Kardashians. What a pity.

Therein lies the hazard of this age group. The electronic revolution which opens the world for so many also creates a monster that actually affects the brain. This is why some psychologists refer to this generation as "brain damaged," unlike any previous generation. Studies measuring cognitive recall of things read electronically versus things read on paper show a remarkable difference. Ditto for notes taken on laptops versus notes taken with pen and paper.

Something about the tactile experience, using more senses in the process, creates a more memorable context. It certainly makes sense, but it doesn't bode well for the functionality of this current youthful generation. When your life and self-concept revolve around "likes" you're far less likely to do something weird and innovative like launching a farm business. "You're what?" I can hear the social media response now.

Further, this electronic culture creates impatience on steroids. All my life, if I encountered something during the day that I wanted to study a little, I'd make a mental note and do some research in the evening when the workday was done. I'd get a book, look at the encyclopedia, ask a neighbor. Now, if I'm in the field with an intern and a topic comes up that requires some further study, out comes the smartphone and a quick Google search reveals more than you ever wanted to know about anything. Immediately.

This is both a blessing and a curse. It gives us instant access to lots of information, but it also addicts us to instant gratification. We can't wait any more. I view this as the Achilles heel of this electronic generation. You young folks jump from one thing to another, expecting instant success. If you're not successful in a few months, you say, "Forget that. I'm not very good at it" and you're off and running to the next thing.

People routinely ask me the secret to my success in a vocation not known for success. My answer is that I'm not that smart; I'm just too stubborn to give up. In his wonderful legacy book *CREATING A FAMILY BUSINESS*, Allan Nation reminds us over and over about what he calls the "slog." No matter how romantic a vocation or business may sound, achieving expertise and success is more a slog than anything else. It means staying in the ring when you'd rather slink back to your corner. It means screaming "I have not yet begun to fight" when your powder is low and the sails are shot to shreds. It means shooting another hour of hoops to become a star.

I hope I live long enough to crank out a book about living on the edge of disaster. Often in my farming career I've felt like I'm careening around a Snuffy Smith comic strip type ledge on a cliff, two wheels on the road and two wheels on the brink of the abyss. More times than not I don't have the answers, but just grip that steering wheel and through sheer determination and persistence, eventually get back on the road. Many times the greatest breakthroughs, the most dynamic innovations, occur in these darkest hours.

I'd like to tell you this farm business thing you're launching will be peaches and cream. It'll be healthy animals, plants, relationships, and bank accounts. Sit down and let the good times roll. I've got news for you. It won't. Oh, of course it'll have wonderful times, but it'll also be fraught with the abyss. And you'll have to cross it. You can't give up. The exit ramp leads to another abyss. If farming is truly in your heart, then stay with it. Divorce is not an option.

Instant anything doesn't last. Only what takes all your soul, blood, sweat, and tears will actually last and leave a legacy. How committed are you? If this passage scares your pants off, good. Better to enter this thing scared than feeling full of testosterone and hubris like you've got the world by the tail. This is the problem with video games: you control things. And if you lose, the game gives you another icon and you go on with the race. Life isn't like that. You don't get fresh icons every few seconds. You can't make it rain in a drought, will the fox into the trap, or dry up the flood.

Your asset is your unbounded physical strength and mental enthusiasm. You aren't jaundiced yet with pain and failure. Good. Leverage your energy. Leverage your life force. Find a mentor and become the best student/servant you could ever be. Learn everything you can now, while you're unattached and lovable, before life bears down on you with responsibilities and baggage. Your greatest liability--youth--is indeed your greatest asset. Ignorance enables you to throw yourself completely into learning and launch. Enjoy it.

What a great gift, to know that you want to farm and what you want to be at the earliest stages of your career. If you truly know that's what you want to do, and you're only 16 or 17, count yourself among the most blessed people on the planet. More and more, people don't come to that place until mid-life. Embrace your heart passion with zest, gusto, enthusiasm. You can afford it; be thankful.

If you're that kind of young person, you'll find older people attracted to you. That smells a lot like opportunity. Respect and honor your elders and you'll be amazed the kinds of partnerships you can create. You might even get a farm inherited to you. Be the first to arrive; the last to leave. Go-getter young people endear themselves to the older generation. Be that young person the older generation can't appreciate enough. If you want to be the beneficiary, you have to indenture yourself to people with benefits. That often means older people.

Yes, them, those older people. But if you're faithful, patient, responsible, and respectful, you might be surprised what falls in your lap. Don't worry about a timetable. Just commit to faithfulness; the rest will come in due time.

Anyone can start a farm business, but each person comes to this venture with assets and liabilities. An honest assessment of where you are in life--experientially, economically, emotionally--will help you chart a pathway in. Once you're committed, don't give up.

Chapter 14

Distractions

Perhaps the most common perception I have when visiting struggling farmers is that they are distracted. Some hobby, or peripheral interest, or nostalgic piece of infrastructure, even a belief, can hamper the farm's efficiency and success. It's hard enough to make a farm fly without a bunch of weights holding it back. My list here is certainly not exhaustive, but I hope it gives an idea of the kinds of things I classify as distractions.

They aren't evil. Most are well intentioned. But they compromise time, focus, and finances, which are too precious to squander on distractions.

1. Bank barns. These iconic structures of the eastern American agrarian landscape fill the pages of coffee table books. Routinely on the front page of rural magazines, these structures were built prior to tractors to facilitate moving materials by hand.

Built into the side of a hill, a bank barn is like a house with two ground floor entrances. The upper side of the hill, or a ramp if the barn was located on flatter ground, led into the main portion of the barn. The bottom floor, or basement, housed livestock. The whole idea was to enable hay and grains to be stored upstairs and then easily shoveled downstairs to the animals.

Distractions

These buildings killed a lot of critters. They are cold, dark, damp, and lack ventilation. Each of these characteristics is unhealthy for animals; taken together, they're devastating. Farmers who grew up around these barns have almost religious devotion to their preservation. But they are hard to maintain and virtually impossible to get into with equipment.

Our farm had one many years ago but fortunately it was torn down prior to our arrival in 1961. The folks who salvaged our pre-1800 American chestnut log cabin house from ruin thought the barn, located 70 yards away, detracted from the beauty of the house. I'll be forever grateful that they tore it down. If they hadn't, we'd still probably be trying to make it work.

Before we had our own band sawmill, Dad and I tore down a couple bank barns to salvage the lumber. For the first twenty years, most of our construction projects utilized this salvaged lumber. I've seen these barns effectively retrofitted for other purposes, from a house to miniature golf to restaurants. That's well and good and a proper use for them. Anything except a barn.

As a barn, though, these structures are obsolete, unhealthy, and expensive to maintain. Unless I had some sort of non-traditional use for it, I would demolish any bank barn on my property and salvage the lumber to build something more functional. You can get twice as much shelter with half the money by re-using the lumber in a pole barn.

Today, demolition crews often pay good money for one of these barns. They come in, take it apart, and resurrect it somewhere else for a house or some other purpose. If you don't have a need for the lumber, or the inclination to tear it down yourself, take advantage of these offers and get rid of the thing.

These barns are not your mother or your child, for crying out loud. Is it really worth going into financial ruin to keep these obsolete icons weather proof? No, of course not. To be sure, I'm glad some rich folks have poured money into shoring up and preserving some of these structures. But how many do we need in order to maintain the memory? How many museums are enough?

Getting all weepy-eyed and clingy to a structure like this can siphon off money and time more importantly spent on other endeavors. Some will accuse me of being ridiculously heartless, a beast toward art and beauty. Sorry, folks. I don't see anything beautiful about pouring time and money into a livestock-killing structure that you can't even get your tractor in just because it's historical and built by craftsmen. Do you really want a successful farm? Okay, then, forget the museum business and get on with the work at hand.

If any infrastructure won't work efficiently, either repurpose it or demolish it and salvage the pieces for something that does. Life is too short to coddle things with high costs every day. If every time you go into it and see things that need fixing, or structural barriers to efficient operation, change the structure. You can't be going into a headache building every day. It'll distract you from getting important work done.

2. Horse. I realize I'm going from the frying pan into the fire on this one because I've already received plenty of criticism from my discussion of horses in *YOU CAN FARM*. This is one of the few things I'm repeating from that book because it's both common and crucial. For the record, I don't hate horses.

Horses are expensive, take a lot of time, and require infrastructure. If you're using horses as draft animals or running a business that makes horse ownership generate income, like trail rides, that's a different story. But too many outfits, especially small ones, siphon off precious resources keeping a horse for themselves or a daughter.

I have no problem with keeping a horse. But don't complain about the farm not being successful when you've got $5,000 a year tied up in a recreational horse. "Well, this is why we bought the land, so we could ride these beautiful animals," whines a misty-eyed horse lover. Fine, but don't ever complain at a farm conference that "this farming is just too hard and doesn't pay and I need government grants to help me stay afloat."

If instead of having a horse I spent $5,000 at Disney World on vacation each year, then complained because I couldn't afford things

I needed on the farm, would you say, "Well, maybe if you didn't blow it at Disney World you'd be able to capitalize some of these projects?" Of course you would, and so would any other reasonable person.

The point is not to hate horses; it's to understand the time and place for distractions. If you're rolling in the money and things are going smoothly and you want to divert your resources to some horse fun, great. I'm all for it. But unless and until you have the farm on sound footing, both in time and money, stay away from a horse. If you get one and don't have time to ride it, you'll start suffering the emotional guilt trip every time you go by and see the unused horse.

Love can turn to resentment pretty fast when things start to pile up. Spare yourself the heartache and financial devastation; forget the horse until your farm is rolling in the money. Then you deserve it.

3. Off-farm recreation. Farming is a lifestyle. It's not a 9-5 deal, a 40-hour work week, a punch in and punch out kind of thing. When people talk about how hard I work, I don't get it. To me, it's just taking care of what needs to be done. I don't consider it hard work.

To me, hard work is getting up before dawn to make the rideshare group or get a parking spot in the car park so I can ride the subway to a cubicle where I meet other people's agendas all day. That sounds like torture to me. Just go ahead and start pulling out my fingernails.

If you want to farm, wonderful. But come to it for love, for passion, for creation stewardship redemption. Our farm work is noble, sacred, and righteous. Certainly non-farm work can be that way too, and I would suggest that people who have that much enjoyment out of their work, wherever it is, do not need as much recreational distraction as those who struggle with meaning in their work. Remember, 80 percent of Americans hate their jobs. That's tragic.

This is no doubt one of the reasons we spend as much as we do on distractive recreation. The Romans called it bread and circuses.

We call it the Kardashians, Las Vegas, and football. I don't want to stray too far here, but this recreation can almost be an addiction if we let it. The farm itself provides plenty of recreational opportunities, from picnics to hikes to fishing, hunting, and camping. Do you know how much people spend for those activities? And we can do it right on the farm for free.

Our children have their own garden or livestock on the farm, not little league sports or ballet. Again, like the horse, I see nothing evil or inherently bad about these other pursuits. All I know is that in my travels, farmers whose lives are in constant turmoil to get Johnny and Jenny to their off-farm activity, have a hard time getting things done and making a profit. If you want to live like townies, go live in town. But if you want the joy and satisfaction of the farm, enjoy the farm and stay home. If you're not creative enough to entertain yourself with all the variety and wonder of a farm, you'll never be creative enough to have a successful farm.

Everywhere I travel people come up and offer me the use of their mountain hideaway chalet or the beachfront cottage when I want to get away. I smile graciously and reply, "Thank you kindly, but when I want to get away, I go home."

4. Heritage genetics. I know I'm throwing bombs at some sacred cows here, but somebody has to say it, so it might as well be me. Am I an enemy of American Livestock Breeds Conservancy and seed banks? Of course not. I deeply appreciate the work of all these genetic conservation groups. But sometimes this issue comes across as cultish.

I've been pilloried and excoriated for using anything other than heritage breeds, as if I'm a pariah among purists. So let's get this straight. My goal is to eliminate factory farming with a credible alternative. As wonderful as it might be, that goal is not going to happen with $150 turkeys that only weigh 12 pounds. Ain't gonna happen.

I'm tired of the whining among the purists that they can't be successful because people won't buy their heritage corn or turkey or whatever. Remember, the more out of ordinary your product and

Distractions

price is, the fewer potential customers you have. The marketing pool is a triangle, with a broad bottom and a narrow top. Everything that reduces the market pool moves you toward that narrowing population. Inconvenience is one factor; price is another.

If you go broke because people don't want a $150 turkey that only weighs 12 pounds and has a razor breast with no white meat, all you've proven is that the world is not ready to accept a $150 turkey that only weighs 12 pounds and has a razor breast with no white meat. You don't need to try it to know that such a critter is a hard thing to sell. Anybody with an ounce of sense can tell you that. So the world isn't as smart as you, or doesn't care as much as you? Cry me a river, we already knew that. You didn't have to go broke to prove it.

Everyman food is still on most menus, and the closer it's priced to normal, the closer it looks to normal, the wider your audience. In taste, texture, and farm ecology it can be different--even *PURPLE COW* different (salute to Seth Godin). That's enough to create differentiation. Marketing is hard enough without arbitrarily cutting out 99.5 percent of everybody due to arbitrary contrariness. I'd rather have a 90 percent perfect product that everyone wants to buy than a 100 percent perfect product that only 1 percent want to buy.

I'm conceding that our chicken and our pigs aren't as perfect as they could be. Sure, we could offer acorn-fattened hogs, or corn-free hogs, but we'd only be able to raise about 20. To cover our living expenses, they'd need to be $5,000 apiece. Who is going to pay that for pork? A handful, yes, but not many. I appreciate the folks who limit themselves to this. That's fine. But realize it's a strategic decision and don't whine about people not buying your product. They're trying to put socks on their feet and gas in the car. If heritage-only is what you want to do, great. But man up and accept the consequences of the decision.

That's the point I'm trying to make. Don't complain that the farm isn't successful when you refuse to produce something most people buy. You can save the world tomorrow. Today you have to stay in business. This is not selling out to the devil; it's pragmatism. It's recognizing a context, and our context in 21st Century America is double breasted turkeys with lots of white meat at three or four dollars a pound. The more you deviate from that, the harder it'll be to find folks who want to buy.

If you don't survive to fight another day, what have you proved? That you're ahead of the world? Big deal. You don't get paid to prove that you're ahead of the world. You get paid to be relevant. That said, on our farm we're trying to create new heritage genetics by line breeding. Rather than stopping at old world genetics, why not add a bunch of options for our grandchildren? That seems like a decent approach.

I know Scottish Highland cattle are cute. But having them in Alabama is nuts. They aren't adapted, comfortable, or appropriate in that climate. Note the name: Scottish Highlander. That's a place. A real place. These cattle were selected for that real place for a real long time. If your place is not like that place, chances are they're out of place. Before you save the world, you have to save your farm. If you don't save your farm, you won't be able to save the world. Plenty of time exists out there to become more perfect.

5. Certifications. In today's opaque and ignorant food climate, fashionable people certify their product with numerous organizations. Third party certification, from organic to animal welfare to wildlife friendly, beckons farmers to fill out paperwork and join for credibility and authenticity.

Farmers who view this as their only avenue of market credibility and consumers who view this as their only avenue for reliability suffer from the same syndrome. It's a victimhood mentality that assumes I can't trust myself to make any decisions so I need someone else to make decisions for me. Just like large food corporations were more than happy to take over domestic culinary responsibilities, plenty of sincere-minded outfits are glad to take over information responsibilities. All for a price, of course.

For the record, I have two underlying problems with the

certification fad. First, these systems incentivize compromise and political maneuvering. When the organization is financed by the people it certifies, the only way it can grow is to get more sign-ups. Watering down standards, or playing politics with big players, both bring in more money and more sign-ups. The whole system is predicated on erosion of standards, and that's exactly why all of them are guilty. Big players get a pass; small players take it on the chin. Over time, the trajectory moves toward easing entry hurdles in order to grow the organization, brand, or movement.

Second, certification assumes we can't trust each other, but is predicated on trust. While many do send out site inspectors, auditing to insure compliance, anyone can clean up for a day. But many rely primarily on checked boxes and paperwork, all of which can be manipulated at least, perjured at worst. At its most basic level, therefore, I see these certifications as philosophically schizophrenic. They assume dishonesty, but rely on honesty for compliance. It's silly.

Beyond these two foundational problems, though, the practical outworking is too much time and money for too little benefit. Further, I would argue that in a day when social media is fundamentally opening accountability, this institutional, industrial-type certification model is becoming obsolete. We could argue that the whole government food inspection program grew out of the opaqueness of an industrial production and processing model. This was unnecessary in yesteryear's village economy when everyone knew who the scofflaws were.

With today's social media via the internet, many believe the village conversation, with its concomitant accountability and openness, is returning. If we follow that argument, it leads directly to the conclusion that the industrial governmental oversight agencies are also obsolete. With drones and the internet, democratization of information yields accountability far better than bureaucratic oversight and paperwork.

If farmers would use these new tools effectively, we can garner story and credibility faster and more precisely than through other organizations. The point is that the new village conversation is opening up confirmation or condemnation quite efficiently. The ability to be confident about picking people and outfits with whom to do business is better than ever. This is the Uberization of the new shared economy. The reason Uber and Airbnb worked was because of the real time accountability between provider and buyer. Absent that, not enough trust would have been possible to make these ideas work. Because both sides can rate each other in real time, the accountability openness creates its own protections.

Getting bogged down in the certification business is generally unnecessary and often a serious distraction. It's not the only way to do business. It's not the only way to get credibility. It's not the only way to get your story out. And too often, these programs are downright squirrely.

For example, animal welfare standards that require 5 square feet per chicken sound great until you find out they eliminate enclosed portable shelters. A farm in Georgia that has routinely panned Polyface for enclosed portable shelters is feeding dozens of eagles with its unprotected free range broilers. Although each of my broilers only get 1.5 square feet of space, they get a new 1.5 square feet every day. The cumulative effect is 60 square feet, which is far more than a bird getting 5 square feet in a free range set-up in which those square feet are the same every day.

These nuances are not trivial. They make a big difference in nutrient density of the chicken because they have everything to do with grass ingestion. Getting smaller increments of pasture fresh every day means the forage does not get stale and the birds eat far more than if they're on a larger area longer. These standards are written by folks who sit in urban offices and don't have a clue about these nuances.

How about another example? Some animal welfare standards require baby pig castration within the first week of life. That's doable if the sow is individually confined and the farmer can reach in and grab out a piggie. But in the most natural farrowing model, several sows get together in a pod and farrow together. These 6-10 sows share nursing and guard duties, resulting in more protected piggies who have a better chance of finding a nursing faucet. The sow with 6 piggies has a couple spare faucets to share with the sow who had 10 piggies. The result is all the piggies do better. But any farmer who tries to go in and swipe a male piggie out of such a pod will quickly be dispatched--I mean assume room temperature, like mafia guys--by these protective sows. In this case, the animal welfare requirements preclude a truly pig-friendly farrowing environment.

Organics is far worse. You can put 15,000 chickens in a house and feed them certified feed from China and get a stamp of approval. But if you run 100 birds on pasture with the neighbor's GMO-free grain, they won't pass. To say that factory farmed organics is superior to pasture-based local flies in the face of science, reason, and spirit. I've elected to just not play these games. Instead of filling out paperwork, I invite customers to come and see the farm.

Industrial organic farms have no-trespassing signs just like their non-certified counterparts. The birds don't see sunshine, pasture, or breathe fresh air. Interestingly, the nutrient profile is identical to their non-certified industrial counterparts. If you want a different chicken, encourage it to eat some grass. That'll change things dramatically.

Farmers who turn back flips trying to comply with every pedantic box on a certification form waste a lot of effort. Put that time and effort in storytelling, videoing, social media, on-farm events, and empirical data. Send your food off to a lab and get a nutrient profile. Check pathogens as well. Be aggressive about explaining your procedures and your philosophy. That will resonate with people just as efficiently as spending days in the office trying to fill out forms and is a better investment for your business.

6. Government agencies. This is similar to the certification discussion above, so I won't belabor it as much. Be assured that you don't need affirmation or help from government officials in order to be successful. Further, I would argue that trying to garner that support actually detracts from getting on with the farm.

Sitting in a government office, entertaining agents, filling out forms--this may appeal to some folks, but not generally entrepreneurial farmers. For the record, I don't think government folks are evil or bad. These agencies have some great folks. But they are great in spite of the bureaucracy, not because of the bureaucracy.

Dangling a cost-share carrot in front of your nose can certainly tempt you to start down this path, but more often than not, it's a side track. Cost share programs don't like portable systems; they like stationary systems with lots of concrete, gravel, and other infrastructure. They like straight fences, not crooked fences. I love crooked fences because the terrain is crooked. In my book *THE SHEER ECSTASY OF BEING A LUNATIC FARMER* I deal with many of these issues head on and in more detail.

Just remember that people who don't need to perform in order to receive a paycheck tend to be less creative than those of us who do. I go into orbit these days when I attend a government-sponsored forage conference and listen to university specialists telling farmers it may not pay to do rotational grazing. At best these agents encourage farmers to move their cows once or twice a week. Aaaargh!

The bottom line is this: If you're getting your information and new ideas from government folks, you're already about a decade behind whatever is on the cutting edge. Government agents can't promote something that isn't double-checked up the wazoo. They can't encourage something that's questionable. And they can't endorse something that's so far out on the lunatic fringe it hasn't entered the consciousness of mainstream orthodoxy. By definition, government ideas have to please the majority of people. Majority endorsement comes a long time--years--after the innovation occurred.

7. Social media. Yes, we're still talking about distractions. I'm a techno-peasant for sure so to even broach this subject is out of my league. I don't even know the terminology. Goodness, it's changing so fast that by the time this book comes to print today's social media platform may be different.

One of the most interesting comments I heard recently at a business conference was this, "If you're not on Twitter, you're making a big mistake. If you're always on Twitter, you're making a bigger mistake." Of course, the room erupted in laughter, but I think this set a good tone for the point. As farm businesses, we have to appreciate the value and power of social media in our world. At the same time, it can suck us in and drain us with distractions.

For sure, successful farmers today use social media, but I've noticed that some who appear the most successful on social media are actually not doing very much but whining about how they can't generate income. The reason is they aren't out there working enough to produce enough to generate the income. They're spending too much time monkeying around on a computer screen. Spreadsheets and YouTube do not actually plant carrots or put hay in the barn.

It truly will be interesting to see how all of this progresses over time. As we reach the electronic satiation point, will our appetite for texts and telling the world in great detail about every time we pick our noses ever diminish? I don't know, but I know that outfits who inundate me with material lose their luster quickly. Communicate, but do so when you have something important to tell.

When the barn is burning down is not the time to be on the internet looking for the best supplier of door hinges. You need to be throwing buckets on the fire. Social media can be as distracting as we let it, and it can be addictive. We now know that social media correspondence gives the same feel-good vibes, physically, as gambling, smoking, and drugs. Social media has a place, but it needs a clearly defined place.

8. Altruistic addictions. This is similar to the heritage genetics discussion earlier, but I want to expand it to all sorts of issues. Stubbornly refusing to budge on inconsequential things can hamper success.

One of the biggest spheres concerns tools and machinery. Like most people, I love museums and exhibits showcasing ancient tools. Hand tools and craft tools truly are amazing. Teresa and I both enjoy seeing the artisans at work in living history museums like Williamsburg. An auger to hollow out the center of a long straight piece of poplar, turning it into a pipe, is beyond amazing. But today anyone with sense would use PVC or plastic. You wouldn't think of spending two days with a hand auger to hollow out a piece of wood in order to move water 10 feet.

But people get hung up on these sorts of things. The technological advancement in power tools--especially cordless tools--is wonderful. I remember when I first began building chicken shelters in my early teens. We didn't have cordless anything. I put them together with flat-head screws and a brace and bit. Today, we use Phillips head dry wall screws and a cordless drill. They don't crack the wood as much and go in about three times easier than the earlier system. And you don't have to drill a body hole.

I, for one, am really thankful that we have these cool tools and machines. Four-wheel drive tractors with front end loaders--wow, what machines! Anyone who knows me knows that my favorite tool is a chainsaw. It's hard for me to imagine life before the chainsaw. A couple of years ago Daniel and Sheri got me a chainsaw history book and my epiphany takeaway from that book was that the chainsaw as we know it has only been available in my lifetime.

Thinking back to that first 50-pound David Bradley from Sears and Roebuck that we used in 1961, these modern powerful, lightweight, skinny-barred saws with kickback protection are truly as amazing as indoor plumbing. I can't imagine not having a chainsaw. Every project would be arduous. Houses would be far more expensive. We couldn't do nearly the silvicultural upgrades we do today.

But too often wanna-be farmers get stuck in some sort of nostalgic worship or hatred of the modern. Often it's born out of strong convictions, like to not use petroleum. Sure, I'd love to not use petroleum. I'd like to not have to pay half my income in taxes. I'd like a society without money. Wampum, anyone? But we're in a real time, a real place, with real needs. While I certainly don't believe every modern invention is necessary or even economical, our decisions on whether or not to embrace something should be based on more intellectual honesty than "I refuse to use electricity."

Distractions

Remember gardening guru Eliot Coleman's response to the folks who chastised him for using propane in his greenhouses: the first need is to stay in business. After that, save the world. We've circled around this argument from many different vantage points, but it's important because people get stuck in their own stubbornness. I'm not suggesting we become spineless followers.

What I'm promoting is a practical appreciation for innovation and a reasoned response to new anything. For sure, we can err the other way. Just because a 60 horsepower tractor is good doesn't mean an 80 is better. Just because a $300 chainsaw is good doesn't mean a $1,000 chainsaw is necessary. But by the same token, just because technology isn't the answer for everything doesn't mean it's not an answer for anything. I've watched many anti-technology folks struggle with success because they're too inefficient to compete in a tool-tech world.

Sure you can build a house without power tools. Sure you can move cows with wooden gate panels. Sure you can dig a ditch with hand tools. But what's the point? Do you want to get the job done, or prove your contrariness? I think building a house without power tools is cool. But don't get so distracted doing that that you don't have time to move the cows or weed the beans. You don't have to be the first to embrace technological advances, but you won't get a medal for being the last, either.

I say this looking in the mirror because I'm still on the flip phone. I know the day will come--perhaps even before this book comes out--when I'll need to upgrade to a smartphone. But I'll probably not be the last guy on the planet to get one. Maybe just next to last. When I feel like not having one is hampering my effectiveness and efficiency, I'll get it. But until then, I'll enjoy being archaic. It gives the millenials something to laugh about, "Wow! I haven't seen one of those in years! Hey everybody, look at Joel's phone!" I'm glad I get to provide amusement at my expense; that's cheap entertainment for the younger crowd.

9. Unscheduled visitors. If your farm is successful, you'll begin attracting attention. Goodness, the interest in integrity farming is so high today that if you just start something you'll attract attention. Friends from college days, extended family, curious neighbors. When you start doing something different, people want to see. And if it smacks of sacredness and healing, they'll come in droves.

They may not buy anything. They may not have a clue about your real situation. But they'll come and interrupt your work, banter, lean up against the pickup truck, anything to get close to this interesting work and person. While the notoriety can be intoxicating and the interest affirming, it can be terribly distracting.

Of all the distractions listed thus far, this is the one with which I struggle most. I love people, stories, conversation. I'm a verbal magnet, the proverbial extrovert. But I've had to take the uncomfortable step of limiting my exposure. It's not easy; it's not fun. But as a farmer, you have to get your work done. Just because you're home at 1 in the afternoon does not mean you're on vacation.

The perception by many people is that when you're home, you're not working. Most people go away to work, to their jobs. The notion that when I'm home I'm working strikes people as odd. You'll have to introduce discipline to keep this distraction from getting out of control.

Many years ago, after spending a week doing nothing but entertaining visitors, we instituted the Lunatic Tour. That was a godsend. We now offer 14-16 of these hay wagon, two-hour tours a season and it funnels people to visit at a certain time. By carving out this devoted time for folks, I can now in good conscience walk by people without stopping to chew the fat.

We use our website and social media to let people know about the tours. Prior to the tours, I felt compelled to stop and talk to everyone who came by. But now I'm not nearly as concerned about appearing discourteous or inhospitable because I've denoted a time, very publicly, when I'll be available to converse. This has allowed us to keep our open door 24/7/365 policy without crowding our real need to get things done.

Distractions

Remember that when folks come to see you, you're the host and they're the guest. That means you can set the rules. If you don't want them to follow you on a certain project, its okay to tell them they can't come. If you don't want them driving around, show them where to park and tell them they have to walk. That'll limit their visit. If you don't want campers, it's okay to refuse them. It's your place and you get to define what fits.

Setting up places where they can enjoy the ambiance of the farm, on their own is a good way to channel this distraction. A gazebo, picnic table, pavilion, porch swing. It doesn't take much to offer inviting hospitality but at the same time maintain enough distance so you can get on with your work day.

10. Failure to cull. One of the foundations of good husbandry is breeding performers and culling failures. The distraction is when farmers get attached to certain plants and animals to the detriment of the farm.

Look, if a large iconic tree is hanging precariously over the equipment shed, take it down before it falls on the building. That just makes sense. I know the tree is cool. I know it had a tire swing on it when you were a kid. I know its silhouette in the setting sun is the stuff of *NATIONAL GEOGRAPHIC* pictorial fortune. But when it's done its time, fulfilled its function, that's enough. Letting it fall on the equipment shed is crazy.

This is particularly true with livestock. That old cow, or old bull, or rooster needs to eventually go where iconic old critters go. Running an animal nursing home may be a nice ministry for a nonprofit, but it sure isn't the way to run a successful farm. Once a plant or animal becomes a liability, it's time to cull and move on. This is not harsh; it's simply the way the world works.

Let's fulfill our stewardship function regarding our livestock and plants. One of the reasons domestic livestock feed us more efficiently than wild animals is that we have this husbandry contract. Domestic animals depend on us to care for them because they are not as skillful to fend for themselves in the wild. In turn, we determine how long this dependency exists and terminate it when our logic deems appropriate.

Turning your cow herd into pets is not the way to have a successful farm. I'm not talking about having gentle, sweet-disposition animals. I'm talking about treating an animal far beyond their vibrantly productive life as if they were our favorite pet. Anyone would put more money and attention into a pet dog or cat than a production animal. But treating production animals like pets gets in the way of profitability and emotional freedom.

Few things demoralize or drag down a farmer more than caring for a sick animal. Anyone who loves their stock appreciates giving second chances. My most memorable was a cow we had. She apparently had a hard birth. I found her half under the electric fence on the edge of a field with her newborn calf walking around looking for something to eat.

The cow couldn't stand; she was paralyzed in her back legs. I got the front end loader, put a chain around her, and carried her to the barn. Once I got her rolled over enough, the newborn calf nursed well. Once that was done, I propped her up the best I could in a comfortable lying position, put water and hay in front of her, and left, expecting her to stand within the next couple of hours. Most of them do. But she didn't.

The next day, I went through the same routine. Then another day. And another day. I felt so sorry for her, but she seemed perky and her calf was doing well. I went to my old-timer neighbor and asked him, "Jim, I've got this paralyzed cow. How long do I give her before I put her down?"

He said, "Give her 30 days. If she doesn't stand on the 30th day, she'll never stand."

I went home and started counting off the days. They came and went: 15, 16, 17, 18, 19, 20. She could scoot just a little with her front legs, but not much. The 30th day was a Sunday. By then I was tired of fooling with her, but decided not to do anything until the next day. I didn't want to do the deed on a Sunday. So we went to church

Distractions

and when we came home, the first thing I saw as I drove in the lane was that cow standing up out at the barn. Day thirty. To the day.

Bless her heart, she did it. She staggered around for a couple of days, but soon her gait improved and she walked well enough to rejoin the herd. She raised that calf to weaning. But we didn't give her a second chance. She went to hamburger heaven at the end of the season.

Culling is one of the most strategic husbandry decisions you can make. Keeping everything performance-oriented maintains a profitable enterprise. Creating a bunch of unproductive pets distracts you from true caretaking responsibilities.

11. PVC Construction. I wish I had a nickel for every collapsed, disintegrated PVC structure I've seen around the world. It's light; it's cheap. But it's not durable for construction. As a conduit for water, PVC is wonderful.

Sunlight gradually breaks down the fibers, making it weak and brittle. Chicken shelters, hoop houses, and lots of other things can be built from PVC, but I've never seen one yet that really stands the test of time. Besides, PVC comes from petroleum, which is nonrenewable. And it doesn't decompose if and when you want to discard it.

I like wood because it stores carbon, is ultimately recyclable, and comes directly from real time sunlight. From a construction standpoint, wood is easier to fix. If it breaks, you can scab on a piece or beef it up with some bolts and screws and be back in business. How do you fix a break in PVC? Or how do you beef up a joint if you realize it's a bit weak?

One of our biggest experimental failures was a portable sheep corral. I had this hare-brained idea of building a 100 ft. X 20 ft. corral out of schedule 80 PVC. That's the gray stuff that's half an inch thick. The nice thing about PVC is that it's slick, so it slides across the ground nicely. We put it all together and hooked it up to the tractor for its maiden voyage. The morning happened to be cold. I pulled it about 10 feet and it shattered.

We spent $3,000 on PVC for that project, plus time building it, and it lasted a total of about 5 seconds. I realized immediately that it would not work so we broke it up and threw the whole thing away. Costly experiment. Trust me, folks, never, never, never build a structure out of PVC. Put it in the ground and run water through it. That's why it was invented.

12. Cheap infrastructure. We all know the phrase "penny wise and pound foolish." It makes more sense if you're British, but most people get the point even where a pound is not a monetary term. This is an interesting balance to the whole chapter about frugality. Yes, I'm all about frugal. But I'm not about cheap.

The difference is that bring frugal is an asset; being cheap is a liability. That's why I titled that chapter "Live Frugally," not "Live Cheap." I like good quality stuff. Things that last always make a better investment than things that don't.

Generally, the market differentiates quality by pricing. I try to never buy the cheapest of anything. Whether it's tractors, chainsaws, or electric fence energizers, you can be sure that the cheapest alternative is just that: cheap. It won't last. By the same token, you don't always need to buy the most expensive; but you for sure don't ever want to buy the cheapest.

We have a couple of cheap post hole diggers we bought from Lowe's years ago. We never use them. They're light as a feather and bend on the first rock. No, you want the heavy duty stuff. When it comes to hand tools and garden tools, you want pig iron. Flimsy tools break as soon as you push them. Buying cheap is a distraction from getting something that will actually function and let you get on with the job. Nobody wins when things bend and break.

If you look at the garden hose section at Lowe's or Home Depot, you'll see half a dozen different brands at different prices. Believe me, you want to go to the expensive end and get the genuine rubber with brass fittings, the hot/cold industrial strength hose. Anything else will deteriorate quickly. Flimsy fittings on the ends can mash and leak if you breathe on them wrong.

The difference between home-owner quality chainsaws and hand tools versus professional quality is palpable. Real heavy duty cordless drills, the kind your professional construction friends use, will always be a better buy than their cheap knock-offs. Most power tools, including chainsaws, have built-in hours-of-use. You can buy power tools made to self destruct after only 10 hours. They're cheap, but people who study such things say that the average cordless drill sold at Wal-Mart only gets used 15 minutes in its existence. You don't have to make a very durable machine for it to last 15 minutes.

Chainsaws sold at large hardware chains are minimal use machines. Many of them are engineered for only 50 hours of use. But if you just have a backyard and an occasional need, that's enough to last a long time. If you're actually farming, cutting firewood and using the saw, you want something designed to run 8 hours a day, day after day after day. That saw will cost three times as much as the cheap ones, but you'll get years of heavy use without any breakdowns. What's that worth?

I'd much rather spend $30,000 for a front end loader, 4-wheel drive name brand tractor with a good nearby service provider than a knock-off at half price where parts and service are 500 miles away. Electric fence energizers are the same way. Only a few brands are worth anything. The cheap one at the hardware store or the farm store is cheap for a reason.

All that said, there is a time and place to invest in better. Farms like ours that decide to be in the meat and poultry business often need refrigeration and freezer space. We had 13 freezers before we finally sprang for a walk-in. Each community has a trader magazine for buying and selling things. Craig's List is also an option. At any rate, used home freezers can be purchased routinely for $50-$100 apiece. You can get a lot of storage for under $1,000 at that rate. Even a tiny walk-in will run you way more than that. But 13 freezers eventually became a logistical nightmare. We upgraded only when the poor-boy model became untenable.

When you finally spring for the walk-in, don't skimp on the compressor. You might buy a used box but put a new compressor in it. Some things have a high risk of failure, and used compressors are notorious. Wait until you're forced into it before making the big investment. Then when you make the big investment, get top of the line so you don't have to worry about it. Otherwise you'll be tinkering around with it every day, and that's a distraction.

These are all distractions. I'm sure I've stepped on some toes here and I don't mean to offend. But these are the things that routinely come up in my discussions. They tend to be emotional things; none rooted in practice and performance. That is why they're hard to prune from our psyche and our farms.

"But I just like that cow," we whimper.

"But I like those feathers on that chicken," we defend.

"But just look at that barn. Isn't it cool?" we crow.

As hard as it is to realize, a successful farm minimizes these distractions and gets on with the work at hand. I wish I could save every pretty tree, every interesting building, every cat that purred in my lap. But life doesn't afford us the privilege of these things anymore than any of us can live forever on this earth or get out of paying taxes. Freeing ourselves from distractions enables us to move ahead with laser focus. Positive traction, unfettered, will get us where we want to go much more efficiently.

Farms are certainly far more than businesses, but they are businesses. Reducing these distractions helps us stay on point, on task, and ultimately displaces these distractions with the joy and fulfillment of success.

Chapter 15

New Opportunities

In *YOU CAN FARM* two decades ago, I did a top picks chapter for agricultural enterprises. Those are still just as valid today as they were then. Pastured poultry is still at the top of the list, and I think even a better choice today than back then. America is eating chicken; it has a fast turnover rate; chickens are child friendly because they're small; enjoys regulatory exemptions from inspection (PL90-492) for market access.

The other enterprises I mentioned at that time were grass based beef and dairy, market gardening and U-pick fruits, home bakery, and sawmill. Of course, I go into detail about why I think those are good picks so won't explain it here. Since that time, however, we've had some real changes in our culture and in demographic trends that give rise to a new set of opportunities. I would stand by my original picks for best farming enterprises. These are simply new opportunities.

1. Agri-tourism. This is events-based on-farm recreation and use. As industrial farms become more hidden from public view, the cultural response is a rising interest in visiting real farms.

Factory farms have no appeal to city folks. They stink; they're ugly; they're not interesting except as a wow factor for machinery and size. But as people yearn for authenticity in a fake world, they crave a true rural experience. Working farms, beautiful farms, interesting farms all find a place of endearment in the hearts and minds of city dwellers.

How many of you have attended a farm wedding or wedding reception hosted on a farm? Obviously to pull this off successfully requires an eye for beauty and a heart for hospitality, but if you're a people lover and you enjoy pretty landscaping, this could be a great fit. Here is a checklist to make this work for you:

- Farm located on asphalt public road
- Farmers must love people and hospitality
- All-weather events structure (refurbished barns are great for this)
- Restrooms
- Changing rooms
- Pond (not a requirement, but it's high on appeal)
- Lighting for after-dark arrivals and departures (you don't want guests stumbling around, falling into the rose bushes)
- Serving counters
- Potable water and sinks (food can be catered, but caterers still need a rudimentary place to work)
- Tables and chairs
- Parking (preferably graveled and lighted)
- Attractive landscaping

All of these characteristics and amenities create an agri-tourism destination for any number of uses. While weddings are the biggest, corporate retreats, anniversaries, birthday parties and seasonal options create a wide portfolio. While it may be hard to imagine why so many people would like to come out to a farm just to have fun, to folks cramped up in the city a nearby pretty farm can be as exciting as Disney--and a lot closer.

Seasonal events include the fairly common pumpkin patch and corn maze, a haunted hay ride in October; sleigh rides (if you have snow) during the winter, and an Easter egg hunt. Capitalizing on festive occasions is a great way to show community spirit and engage with seasonal good will in a marketing campaign. Offering on-farm dinners can value add your products to retail plate price; don't forget to double the meal price for the unique ambiance a farm offers to the gastronomical experience.

Some farms now offer movie night once a month. Think about what you can offer that folks can't get anywhere else. At our farm, we made a fire pit and offer hot dog roasts for groups. Do you know how many teenagers have never cooked a hot dog over a fire? Or never had a s'more? They hear their parents talk about doing this on youth group outings, or with Boy Scouts, but today's teens often never personally experience any of this. It's all banished to the historical narrative of yesteryear.

I'm reminded of the guy who told me his city cousins' favorite activity--and later their favorite memory--was "hog fishing." When the city cousins came, the kids would all go to the woods and select long, skinny tree sprouts. They cut them down with hacksaws and cleaned off all the branches. With a piece of baler twine tied around the end of the pole and the other end of the twine tied around an ear of corn, these youngsters would sit on the hog yard fence and dangle their bait out over the pigs.

The pigs would come up dutifully, smell the corn, and bite at the cobs. "I've got a big 'un! I've got a big 'un!" Gleefully, the youngsters would tease the pigs with the corn and when a pig latched onto an ear, it would pull back just like a fish. This tugging, with child on one end of the pole and pig on the other, offered many a fun time for these visiting city kids. They all called it "hog fishin'."

I've met farmers who were really actors dressed up like farmers. Attend any North American Farmers' Direct Marketing Association (NAFDMA) conference and you'll see more fun things to do on farms than you can imagine. Elaborate theater shows, hog races, pony rides, corn boxes, pumpkin tosses. You can't imagine all the creative ways these entertainer farmers get people to come out and drop money at the farm gate. One farmer I know purchased half a dozen peddle tractors and installed a nice crushed stone riding arena surrounded by a log boundary. Kids go crazy to come out to the farm and ride those tractors. In case you didn't know, where kids want to go, there goes mama--with her credit cards.

One fellow built miniature golf holes on pieces of plywood so he could reconfigure the course wherever he wanted just by picking up the stations with forks on his tractor loader. With the demise of the circus, the demand for this kind of on-farm entertainment with real animals, real plants, and real country sites and smells, will only escalate.

2. Edu-tainment. I'm making this a different category, although it's a close cousin to agri-tourism. The reason it deserves a different discussion is because the primary goal is education rather than recreation. While agri-tourism can be defined broadly enough to include this, I think it's important to point out some critical differences.

Visitor expectations vary widely. On the recreational side, landscaping, bathrooms, and facilities are paramount. But with edu-tainment, the most critical expectation is that folks leave having learned something. That means the farm can be far more rustic. Patrons are more willing to put up with rougher conditions for edu-tainment. If they really learned something on their outing, folks will forgive the landscaping, slogging through some mud, or whatever.

Further, this enterprise appeals more to farmers with message and method rather than strictly theatrical prowess. You don't have to be a showman to pull this off, but a bit of personal flare certainly makes it easier. As Richard Louv, who coined the phrase "nature deficit disorder" points out, the need to connect with our ecological umbilical is imbedded deep within the human psyche. In an asphalt world with nothing but video game entertainment, communing with something meaningful on a farm is a salve to the spirit.

A local naturalist could lead hikes on your farm. From edible wild plants to bird watching to identifying tracks and insects, farm as classroom is a burgeoning opportunity in farming. This certainly may not be your primary income, but it can be a healthy side business. Unlike most farm enterprises, this one does not require large infrastructure overheads.

Informational tours around the farm, showing what you do, why you do it, how you do it--just your everyday stuff--is fascinating to someone who has never seen it or done it. What seems hum drum to you is magical and captivating to others. Don't sell your knowledge or skill short. Classes and seminars on survival, primitive artisanship from tanning hides to hand-tools, cookery and construction are exploding today. Why not host them on your farm?

Find the gurus for the different skills and topics and invite them to conduct a seminar at your farm, using your wood, fields, clay or whatever as working material. That makes the educational environment live for attendees.

Don't forget basic ecology tours. Our farm hosts thousands of children a year in what we call Grasstains tours. These are two-hour walking tours around the farm where we explain the ecological justification behind why we do what we do. Yes, groups pay for these things just like they would if they went to a museum or zoo. Dovetail this with a marketing effort and you get a two-fer. All these kids should go home with a brochure and an introductory $10 off coupon.

Link up with eco-tourism in your region. Both public and private agencies offer hundreds of bus tour opportunities to visit interesting places. Why can't your farm be an interesting place? Put together an offer as part of a package, or offer a package just for your farm. Throw in a lunch of brats on a wood-fired griddle, or pizza from a masonry oven, and you've got the makings of a memory and experience no one will forget.

The key here is that you either have to be an expert, or know an expert who will lead it. These types of experiences have savvy, discriminating participants and they'll sniff out a charlatan quickly. If you're going to offer cheese classes or gardening, make sure the teacher knows what she's doing. In a hurried and harried world,

New Opportunities

people love to combine education and entertainment; our farms are one of the best places for such a confluence.

3. Urban custom farmers. I think the opportunity for urban farming in general and custom farming specifically is wide open. Urban farming has many permutations and plenty of material exists on how to do them so I won't belabor it here. One of the most interesting urban farms I've ever visited is the Brooklyn Grange up on top of the Naval Yard roof in New York City. And yes, they have multiple income streams, two of which are agri-tourism and educational programs.

These operations offer unique challenges but also unique assets. For a dirt road rural hillbilly like me, the thought of being surrounded by customers within walking distance makes me drool. But it's not all roses. Regulations, theft, difficulty finding carbon and manure--all of these make the urban sector challenging.

Custom farming in the urban sector, though, offers some unique positives. Many urban foodies with back yards would love to see something edible growing there, but have neither the knowledge nor time to do it--or at least they think they don't. Similar in form to custom chefs who fix meals in-home, these custom urban farmers ply the neighborhood with their gardening tools and turn yards into food meccas.

Generally, the home owner uses the food. The farmer charges by the square foot rather than by production. A homeowner has a plumber, electrician, and repairman; why not add a farmer? The beauty of this arrangement is that the farmer does not have to juggle production, inventory, and marketing. The farmer simply fills the agreed-upon square footage with produce and the host family uses it. I think this is one of the coolest ideas to come around. When the urban family adds "my farmer" to their lexicon of service providers, like "my pediatrician," "my investment advisor," and "my yoga instructor," you'll know the civilization is turning a corner toward remediation.

The farmer need not own any land, so capital entry is low.

Generally it's all done with hand tools. The biggest tool may be a broadfork, but everything else is small and can fit on a bicycle or motorcycle. At most, a tiny vehicle. This idea can extend to yardscaping at businesses, including condominiums. Why not turn the yard into production? How about the local hospital lawn? Convert it to edible landscaping and garden beds. Many institutions have expansive lawns and landscaping budgets. Why not convert those into real production?

People who argue with me about feeding the world with local integrity food don't realize that in the U.S. alone we have 35 million acres of lawns and 36 million acres housing and feeding recreational horses. According to production figures compiled by gardening guru John Jeavons, that 71 million acres could feed the entire country without a single farm. The cost of maintaining edible landscaping rather than strictly ornamental landscaping is about the same; why not have it edible?

This is convenience farming just like we have convenience food because so few people cook anymore. With 30 percent of all food consumed in automobiles and more than half of all food prepared outside the home, the convenience farm seems like the most logical next permutation of the trajectory. I think this is a golden opportunity that more wanna-be produce farmers should pursue. Why not make the custom backyard farm the new status symbol for environmentalism and food activism? We should start a new movement and take over the yards of urban America.

4. Agri-community. Another fairly recent development is the option to keep farms operating by dovetailing them with an on-site community. As of this writing, some 200 of these are ongoing in the U.S. They're in various stages of build out. They come in many permutations, but the basic scheme is a housing development tied by covenant to a farm nucleus.

Often they contain some sort of co-housing arrangement wherein the development shares septic, water, and energy. It could be solar panels or it could be a central wood-fired furnace that heats several houses. The idea is that sharing these large infrastructure expenditures lowers the cost on each house. Instead of each dwelling needing a separate well and septic system, for example, one big installation services everyone.

Earth ship villages are an extreme example of this. I've been to one in the Netherlands a couple of times and it's wonderful. A created wetland purifies all the sewage. All the houses have an earth bank berm with living roofs. They used all recycled products, including billboards as interior paneling. Super insulation and all the earth sheltering keeps everything warm in winter and cool in summer, further reducing energy needs.

As the negatives of completely independent living escalate, these tight community arrangements become more popular. The sheer expense of maintaining costly infrastructure spread far apart becomes a drag on a culture. These tighter knit options have merit by reducing resource use and financial outlay. They have an advantage in offering guaranteed community. People who live there do so precisely because they do not want to live in a place where they don't know the folks across the hall. In many ways, it re-creates the village concept. Over the course of human history, more people have lived in villages than in cities or even rural areas. It's a time-tested model and one that undoubtedly will rebound in coming years.

Residents of an agri-community generally have a community-centric view toward living and believe mutual interdependence is better than rigid independence. This scheme came about as a way to save farmland from being chopped up into residential estates and scattered development.

Zoning policies need to be modified to enable these kinds of developments because I deeply appreciate the notion of connecting a community emotionally, logistically, and financially to a farm. This benefits both residents and farmer. As someone who has spent countless years peddling my stuff to town, the thought of having 30 or 40 families living on the peripheries of my farm paying an annual food fee, like a glorified CSA, intrigues me.

While having a bunch of people around would make many farmers cringe--since most farmers don't like people--to others it's a godsend. Some of these agri-communities are popping up around hospitals or colleges. Why not tie a retirement community to land adjacent to a hospital or already owned by a college? Or both? A built-in customer base releases the farmer from pavement-pounding and it gives the residents a secure source of food.

If the teamsters strike or the longshoremen walk off the job, residents can see their food growing or walking around in the fields. These models attract people who want to live proximate to their food, want a connection to a farmer, are tired of the fragility and vulnerability of supermarket food, and who enjoy a home-centric food system. They aren't afraid of their kitchens or some dirt on a tomato. A farmer surrounded by participating, active cheerleaders receives both emotional and financial support.

The problem, of course, is when some militant vegan moves in and complains about the chickens. Or perhaps some heritage breed purist moves in and complains about the Better Boy tomatoes. Or one family has hellions for kids who run rampant through the green beans, or leave the gate open to the hog training pen. We can all imagine a host of nightmares.

But being out on your own and drumming up business has some nightmares too. No paradise exists this side of eternity. The agri-community farm is an interesting concept that is gaining traction. I think as more of these projects come to fruition, they'll stimulate more farmers and investors to get together and make them happen. The momentum is there. The positives are myriad.

For the most part, what's been holding them back is local zoning ordinances and regulators. Because these are new ideas and new ways to use land, they don't fall neatly into conventional development boxes. It's a Home Owners Association surrounding or integrating a farm. When most HOAs prohibit clotheslines and chickens, this is truly a different critter to government agents. But with each approval comes more precedent, more creativity. The ones that are up and running offer new experiences to incorporate into future projects. All of this will make the next generation of the agri-community more viable.

New Opportunities

5. Therapy farms. I've been on a few of these and seldom leave before shedding some tears. The spirit of the farmers in these situations is the same as the nurses who clean bed pans in nursing homes. When I think of angels here on earth, these people always come to mind.

A family stopped in here at the farm a couple of days ago that had a huge affect on me. They arrived in a van and wanted to look around. Sure. As they began getting out of the van, I could see that it was an unusual assortment of people. Latinos, Orientals, whites, blacks, and all but two were blind; one was in a wheelchair. Turns out Mom and Dad had adopted--ADOPTED--nine special needs children from all over the world. I just began to weep. Who has a heart that big? A very, very special person, that's who. My problems pale into nothingness next to a family like this. God bless them.

In the same vein, a spirit of patience and care in the face of revolting conditions makes institutional caregivers deserve a pedestal in my book. We've all been in institutional settings where these caregivers labor, day after day, trying to bring respect and honor in a world of walls, bed linens, wheelchairs and sterility.

The first therapy farm I ever visited was for brain-injured people. The farm makes cheese and has a wonderful bakery. Residents--you could call them patients, but most of these places call them residents--do most of the work. They're supervised by a dedicated, loving staff. Obviously higher functioning residents do more demanding work and the lowest functioning the lower. But each resident spends the day doing something meaningful and honorable.

For the least lovable of society, can you imagine anything more honoring than farm work? The physical labor, much of it repetitive, is perfectly suited for these disadvantaged folks. In doing this work, they are validated as worthwhile people, praised when they do well, and routinely encouraged. Can anyone imagine a better environment to develop to as high a functional level as possible?

One of the most interesting therapy farms I've been on was one dedicated to autistic adults who had aged out of the public school system. Rather than simply institutionalizing these folks, their families decided to place them on a therapy farm that offered chores. The residents cared for a handful of animals but spent most of their time in greenhouses and gardens. Located next to a heavily traveled road, the farm had a produce stand where people could stop in and buy stuff.

This produce stand created a sensible and sensitive interface with the residents and their customers. For these autistic adults, many of whom had suffered countless emotional abuses from schoolmates and others, to receive the smiles and praise from customers was life changing. The farm had a special trail designated as a "tension release" system. Rather than administering drugs or using restraints when a resident "acted out," the staff used this designated path as a decompression option. It worked beautifully.

By the time an out-of-control resident walked--or ran--down the path, through the woods, around the chicken yard, and across the orchard, he calmed down and went right back to work. The idea worked so well that once the residents had been there awhile and grown accustomed to protocols, they would self-medicate with the path when they felt an aggravation tantrum coming on. One fellow had massive disfiguring cuts around his face and arms. I asked the staff what happened, and they said he had jumped through a couple of plate glass windows in autistic acting out episodes, but here on the farm, he could blow his stress off by zooming around the decompression path. Isn't that cool?

A close cousin to these farms are the ones being developed by the Wounded Warrior program. From Post Traumatic Stress Disorder (PTSD) to catastrophic physical dismemberment, more and more farms are popping up to serve these American servicemen and women. With raised beds and some design adjustment, you'd be surprised what a mobility-disadvantaged person can do.

One of my dearest friends and mentors back in the 1970s was a paraplegic who was wounded in WWII. Although wheelchair bound, he did more farming than most able-bodied folks. He built garden beds high enough that when he rolled by them in his wheelchair the soil was at about eye level. He kept one of the most beautiful, organized small farms I've ever seen, all from a wheelchair and motorized equipment.

For farmers who have a heart for this kind of ministry, I think the therapy farm is a wonderful opportunity. As the director of the autistic farm mentioned above told me, "We're selling medical care. The food is ancillary."

I don't want to trivialize the farm aspect of these models because without the farm, they're just another institution. But think about the financial security of a farm selling care. You're not subject to crop failure; you don't have to beat the bushes to drum up customers; you don't have to watch your margins as closely. Again, as someone who has wrestled with tight margins and earned my living by what I could produce and peddle, the idea of a steady monthly income from a private or public medical insurance stream whether the farm production is successful or not has a definite appeal.

For those who think I'm straying clear off the reservation calling something a farm that doesn't deserve to be a farm, I would simply respond, "Can you think of anything better for a farm to sell than healing? Why not sell care? Is care less valuable than a carrot? Is therapy less important than thyme? Really?"

You see, I think our idea of farming needs to expand. We've suffered long enough under the industrial assault, now codified into too many zoning ordinances, that a farm is simply a producer of raw commodities to be value added by smarter and wealthier people somewhere else. This is why it's illegal for us to use our sawmill to cut a neighbor's log into lumber. This is why it's illegal for us to make a chair out of our lumber from our own tree on our own farm and sell it to a friend.

The segregation of the farm from the rest of society is part and parcel of the problem. Agriculture zones are so narrowly defined that all the historical manufacturing and value adding that used to be done on farms is now illegal. You can't even have a wedding or birthday party some places without a permit. It's time to recognize the multi-dimensional ministry and function of a farm.

My mother's best childhood friend was the daughter of the head farmer at the County Home. This was the nursing home for the indigent. They milked cows and tended gardens. It was essentially a self-contained farm. The therapeutic benefits residents enjoyed from being surrounded by life abundance is unquantifiable. For sure, a stinky factory farm would not have the same affect, but I don't think anybody believes my therapy farm idea includes a walk through the factory chicken house.

Mounting evidence indicates dietary relationships with autism and other maladies. The rise of food allergies has come right along with GMOs. Isn't it a perfect example of nature's accounting method that the best place to receive therapy for these problems is a farm? A place where nutrient dense food, compost piles, and functional abundance drives the economy. Back to the land, back to the soil, back to real food--that's the path toward therapy.

I salute any farmer who takes on this mission. I think it's sacred, righteous, and compelling. The trend line indicates that over the next couple of decades we'll see an increasing population that needs these services. Farm and farmer as healers is as high a calling as anyone could request. Just like I'd like to see factory farms vacant, I'd love to see institutional living shift to the vibrancy of farm living wherever possible. Who will step up to this challenge?

7. Elder-care farms. This is a bit of a spinoff of the therapy farm, but again, I think the differences are significant enough and the opportunity great enough to create a different category. When anyone asks me to identify the worst served segment of our population, I don't hesitate: the elderly. No civilization has despised its elders like modern America.

It starts early, and like most things, the elderly have brought it upon themselves. Parents dutifully send their kids off to school to be educated and molded by a surrogate in order to regurgitate a body of information well enough to get good grades so they can get into a college at a high enough achievement to land a high paying job in a city far, far away so they can earn enough money to put mom and dad in a nursing home.

And what do mom and dad get in the nursing home? They get the cheapest junk food you can imagine, while paying $500 a day for the privilege. One of the most frustrating conversations I ever had was with the administrator of a modest 40-bed nursing home here in our county. I had the meeting at the behest of a new resident who had been a longtime Polyface customer. As a faithful patron, this lady knew our food hearkened back to what she ate growing up on a farm.

She didn't need young foodies to tell her that compost-grown tomatoes were better than chemicalized and that pastured chicken was better than factory. When her family put her in the home, she marched right into the administrator and demanded Polyface food. She caused such a ruckus that--perhaps in self-defense--the administrator arranged a meeting with me.

We exchanged the normal niceties and then got down to business. Before too long, I found myself asking her, "Where do you get your chicken?"

Her response is one for the classics, "We wait until it's on sale and buy the cheapest stuff we can find, as close to sell-by as we can get. We only buy it when it's at salvage prices."

I was speechless. Here she was charging residents $500 a day but couldn't abide an extra dollar to actually put good food in front of them. For snacks the residents had their choice of Triscuits or Cheerios. Wow, what a choice. At $500 a day!

I had the audacity to tell the administrator, "You could probably spend less time marketing your beds if you bought good, local food and made that part of your care statement. Imagine if you told families 'we'll love your mom so much that we'll feed her the top local nutrient-dense pathogen-free food we can find.'"

She looked at me as if I were from Pluto. We were that far apart in paradigm. After the short conference, I had to walk back to our patron and explain the conversation to this dear lady. She deserved better. But she got junk.

This dearness to my heart dates way back. Our farm has a high hill overlooking the farmstead. The high bluff is across the road and kind of off by itself. My dad always dreamed of having a retirement home up there, surrounded by gardens, where residents could piddle around the garden pulling weeds and grazing on vegetables. The more ambulatory ones could walk out through the fields, commune with the cows when they were close, and come down to see what we were doing.

We had an elderly neighbor whose family took care of him until he passed away. He lost his eyesight and some toes due to progressive diabetes, but never lost his sense of humor or his mind. His family would drive him over to our farm (about 2 miles away) when we dressed chickens. He'd sit in a chair as close as he could get without getting splashed.

He looked forward to those visits more than kids anticipate going to Disney World. He could hear and smell everything that brought a lifetime of pleasant memories to his mind. He would tell us stories and we'd banter back and forth. It was one of the most delightful interactions we could imagine and although he passed away twenty years ago, I still miss those times. I know those visits brought unparalleled joy to him in his waning years.

Of course, Dad's idea would be virtually impossible today because now we have agriculture zoning. You might as well ask to build a nuclear reactor as ask for a special use permit to put in a ten-bed elder care facility. But in other areas this is possible. Farms that happen to lie in commercial zones are perfect candidates for this. It's a way to preserve the farm land, create a couple of farmers, and do some real good for the residents. The farm could even supply the food for the facility. I believe a place exists for many different kinds of farms than what we currently have. Why not have some integrated emotion, economy, and ecology? Who better to enjoy it than our elderly in their final years?

8. School farms. While I'm certainly a big fan of school gardens, I'm an even bigger fan of school farms. Why stop the creativity with a garden? Why not extend the creativity all the way through the curriculum, the buildings, the whole shebang?

As public education becomes less desirable, interest in alternative teaching models increases. Of course, Waldorf school and Montessori schools have incorporated a lot of craft, nature, and gardening in their curriculums for a long time. But this doesn't go as far as school farms.

The first school farm I ever visited had a profound impact on me and I've never forgotten it. It was Camino de Paz in New Mexico, a farm school offering a three-year program for 7th-9th grade. Rather than having experiental farm learning as an adjunct to the curriculum, the entire structure of the school actually served the farm. Although it was not a boarding school, the environment had a life-altering affect on the students.

What I remember most was that as I greeted the students, each one looked me in the eye, offered a firm handshake, smiled politely, and vocalized a mature introduction. I was blown away by the maturity and demeanor of each student. These youngsters took me on a tour around the farm--only 5 or 6 acres--explaining their methodology, asking questions for advice, and discussing protocols among themselves.

Although it's been many years since I visited, my fondness for the spirit I encountered there is as clear as it was the day I left. They had a herd of milking goats. Yes, the youngsters milked them and made cheese. They analyzed the milk for bacteria and kept all the production records. A couple of greenhouses complemented the expansive gardens.

They sheared the sheep, carded the wool, made thread and then handmade computer bags and smartphone holders. They made all their meals, from bread to ferments to meat. They canned, dehydrated, and froze food. All of these activities provided plenty of math, science, history, and language arts lessons. The farm did not supplement the learning; the learning supplemented the farm. These students had a vendor booth down at the local farmer's market where they sold their produce and wares. They made soap and other crafts. They spun honey. They butchered the animals. They ran gross margins, balance sheets, and learned accounting through the various enterprises. They learned sales, marketing, signage, messaging as they peddled their products. And these youngsters were 13-15 years old!

Tragically, the whole operation struggled financially because, as the directors said, "Parents who send their kids to private schools generally don't look kindly on farm work." My question is what parent wouldn't want their kids to have this experience? While I was there, the school had a banquet and the students served the meal. I was moved to tears at how these 13-15 year-olds handled themselves. Typical of my alternative school encounters, my reaction was this: what a tragedy that every child doesn't have an experience like this.

Permutations on this theme are summer learning camps that offer similar experiences but in the off-school year. I was on one of these farms on Martha's Vineyard many years ago. The children moved chickens, then dressed the chickens, then ate them at a local restaurant after a chef cooked them. Believe me, these children were not emotionally disturbed by the experience. They demonstrated an amazing sense of wonder, accomplishment, and self-awareness about the whole process, and about life. They encountered death, sickness, sales frustration, manufacturing details--who *wouldn't* covet these opportunities for their children? Kudos to their parents who signed their waivers and sent them to such a maturing experience.

The school farm is a perfect opportunity for educators and farmers to team up and do something innovative and culturally healing. The chance of having one person gifted enough in different areas to pull this off is slim to none. But two people, working together, with an MOU and clearly defined spheres of responsibility, could make this opportunity fly. When you realize that Bill Cody was only 13 years old when he rode for the Pony Express, you realize how we've undersold and under-affirmed our young people. It's time for adults to step up to our responsibility by creating new places to flourish for our youngsters.

9. Camp farms. This is a bit of a call out to Boy Scouts, religious camps, history camps, nature camps, and others to incorporate integrated food production with the greater organization. Most camping outfits have lots of land and spend countless hours and dollars mowing and landscaping.

Few outfits could be more perfectly suited for synergy than camps and farms. Unlike colleges, summer camping programs always synchronize attendance with the growing season. Seldom does Minnesota offer day camps for kids in the dead of winter. A natural symbiosis exists on the timing, which is half the battle for integrated food systems. The food production timetable needs to correspond with the instructional/recreational timetable. The people need to be there when the production is there.

Typically camping programs are run by people who think only about camping programs. I challenge these administrators and staff to think about incorporating farm production either into the program or as a separate land use function. We have a Boy Scouts of America camp about one mile as the crow flies from our farm. For several years they had an instructor (these are usually college-aged young people) who brought scouts over for a farm tour each week so they could receive their "Sustainability" merit badge.

How much better would it have been if these scouts had actually planted, weeded, harvested, and prepared something over at the camp? It's not like they have no room over there. They have fields and forests--lots and lots of areas where active production could occur. Certainly a small chicken operation could be installed to handle all the kitchen wastes. I guarantee you these urban scouts would fixate to the chickens in a heartbeat. It would be the most popular spot in the whole camp.

The Garden Clubs of Virginia have a wonderful nature camp in our county. It's been operating for decades and the youngsters who attend study almost college level biology. It's impressive. But the food is industrial junk. For a couple of years they had some staff who insisted that they use our food. They brought the campers over to our farm for tours and then the kids ate our eggs and sausage for breakfast. It was a wonderful connection, a closing of the loop, and made a practical lesson. Rather than communing with nature "out there" they were able in this way to internalize the whole sustainability equation.

Who takes the lead on this camp farm idea? Is it the camp administration or a neighboring farmer? I have no clue. But I think a young person who wants to farm and camp administrators who want to ratchet up the camp experience (and marketability) need to appreciate the synergy of this idea. Let's get this done. It's way past due.

I'm convinced that if we will let our imaginations run, we can see how farms, in all their manifestations, can minister to the needs of body, mind and soul. Appreciating all these niches provides a richness to the farming world. When we say we're farmers, it's more than "plows, sows, and cows," as many 4-Hers point out. Farmers come in all shapes and sizes, all kinds of interests, talents and gifts. A farm fit exists out there for anyone. Now go find it.

Summary

Thank you for sticking with me through these pages. We've covered a lot of ground and I hope by now you're itching to improve your current farm or ready to wade into the farm of your dreams. In either case, renewed enthusiasm as a farmer is always a good thing.

Remember, I consider this the graduate course following the introductory *YOU CAN FARM*. I've spent some time with it at my side during this writing process to make sure I didn't repeat things. I have purposely left out some foundational things in order not to repeat the material in that book. At the risk of sounding like a huckster, read that book if you haven't. It'll make this one seem even better.

I now have 50 years of active memorable farming experience (we won't count the first 10 years of my life). That means I'm entering the old geezer phase. I've been remarkably blessed to be on countless farms both in this country and around the world. I'd still rather visit a farmer than Disney World. I haven't been there and have no desire to go.

I'm such a farm boy at heart that I scare myself sometimes at how out of touch I am with pop culture . . . and how little I care.

Summary

But I'm hopelessly in love with healing the land, serving enthusiastic clients with honest nutrient dense foods, and promoting a farm and food ethic that honors farmers. I believe that how a culture treats its farmers, so goes that culture. To go one step further, as a culture treats its soil, so goes the culture.

Farmers have the most common sense, the most patience, and the most hopefulness of any group of business people. But we farmers can be the most pigheaded, silly, unbusinesslike people too. I certainly don't have all the answers and never will. The day I quit learning is the day I want to quit breathing. I hope during this visit together you've had the discernment to hold onto the grain and spit out the chaff.

When I wrote *YOU CAN FARM,* I tried to put down everything I thought was important. I could not have imagined that in 20 short years I would learn nearly that much again. Nothing in that book is wrong; in fact, it's really good. But what I know now is more refined; it's more mature; it's simply more.

As we sail into the internet age, I'm seeing more and more young farmers passing themselves off as experts. They're e-savvy and can set up blogs, podcasts, virtual classrooms, and video-conferencing fairly efficiently. I'm frankly quite concerned about this half-baked advice coming from 30-year-olds into the hearts of their peers. I've read, listened to, and watched some of this material, and it frightens me. It's not time tested, tried and true.

Just because a person can whip up a sharp e-teaching platform does not mean he's an expert. My material may not be glitzy. I don't even like doing video teaching in a sterile studio. I'm amazed at how fast people will follow some tech-savvy advisor who's been in business a couple of years, supported with off-farm income. Gentle people, we're as responsible for the mentors we choose as the advisors are for what they teach.

Let this be a word to the wise: beware these tech-savvy advisors who, with glitz and smartphones wow you into discipleship with electronic pizzazz. The folks who actually do things rarely promote themselves. The folks who actually do things are too busy performing the visceral, real tasks to stop and e-blast everything in their lives. Be discerning. Be astute. Look at accomplishment and time. Look at what's real. Don't be taken in by flashy instructors. Look for the old-timers, the ones who have been in the trenches for decades, the elders. That's where the gold lies. I've tried to acquaint you with a lot of these folks in this book. Listen to them. They are oracles of truth.

The advantage to being a truly callous-handed practicing farmer is that my advice grows directly out of a life in the trenches. My paycheck doesn't come from off-farm investments or old family money. I could go belly up tomorrow with only a couple of ill-informed decisions. It wouldn't take much. Skating this close to the edge all the time doesn't make me fearful, but it does make me watchful.

I'm confident some people are offended at some of my advice. I appreciate that. But I'd rather speak the truth as I see it than remain silent in the face of irritation or opposition. I'm sure something in this book is wrong, but not much. If I didn't believe it, I wouldn't say it. My only motive is to encourage and facilitate a new generation of land caressing entrepreneurial transparent farmers to launch businesses and service their communities on the millions of acres that will come available over the next 20 years.

I haven't arrived and neither have you. We adjust; we cogitate; we develop. And isn't that the way life is. You don't arrive; you just keep navigating. And you navigate. Then you navigate some more. A successful farm expresses this never-ending quest toward better.

My prayer is that these principles will propel you onward in your journey, regardless of where you are. Goodness, I need to look at these principles too, to be reminded that I still stray off course and need to make corrections. Truth is a wonderful thing. It points the same direction. It doesn't waver.

As I've navigated this half-century of active farming, I think I've found some truth. Not all of it, but some. My hope is that this truth will set you free to navigate your own journey. Farms

Summary

are as unique as their farmers. But the similarities among the ones who thrive are as consistent as the similarities among the ones who flounder. With all my heart I want to see an America filled to overflowing with thriving, vibrant, multi-generational, soil-building, water-holding, air-purifying profitable farms.

I'm testament to the notion that we can have it all: production, profit, and pleasure. I trust you'll find that in your successful farm business.

7 Habits of Highly Effective People 89

abandonment 4-5, 7-9, 16
abattoirs 158, 170, 216
Ableman, Michael 9, 33, 205, 215
Access 11-14
accounting 168-185, 232-235, 300
acquiring land 66-67
Acres USA 23
aging farmers 26, 44, 67, 77, 202, 203, 248-253, 296-298
agri-community 290-292
Agri-Dynamics 187
agri-tourism 284-287
altered landscapes 7-9
Amazon 135-139
apprenticeships 63, 254-257
ATV 52, 53
autism 296

banks barns 262-264
beaver ponds 187-188
benchmarks 231-239, 245-247
Berry, Wendell 33, 217
big-supermarkets 126-130
broad-based dips 11
broilers chickens 54, 110-111, 165, 213-214, 215, 232
Brown, Tom 34
Brunetti, Jerry 187
Buffet, Warren 87-88
buildings
.... 10, 18, 26, 41, 54, 66, 74, 79, 87, 175, 203, 206, 209-210, 262-263

CAFO (Concentrated Animal Feeding Operation) .... see factory farming
camp farms 301

Index

Carbon	20, 154, 177, 187, 188
careers	8-9, 10, 62, 248-256, 265
caregivers	8, 293-298
cash flow	164-165, 166
certifications	268-272
change	3, 7-8, 16, 18-20, 56-58, 253-256
cheap vs. frugal	280-282
chefs	139-141, 162
chemical fertilizer	41, 177
chick brooders	114, 211, 224, 232
chickens, processing	112, 224, 229-230, 233
City Chicks	236
clumping/doing like things	239-242
Coleman, Eliot	32, 102, 205, 212, 229, 236, 275
college debt	see educational debt
commission based pay	83-84, 87-89
Community Supported Agriculture (CSAs)	134, 141-143
compatibility issues	78-84
compost/composting	6, 48, 195, 242, 296
compound labor	46
Covey, Steven	89
cow-days	172-177
culinary arts, loss of	93-94
culling, importance of	277-279
Curb Market stand	42
debate	56-58
debt	36-37
debt free	41-42, 45, 67, 254-255
delivery	159-161
depreciable infrastructure	48
desertification	6
direct marketing	92-107, 150-167
direct selling	see Metropolitan Buying Clubs

309

distractions 262-282, 230-231
Doherty, Darren 12, 179
domicile 10-11, 44-47
Drucker learning curve 59

earth ship villages 291
earthworms 79, 177, 199, 200, 205
ecology 4, 9, 15, 200
ecological degradation 5
edu-tainment 126, 144, 287-289
educational debt 254, 256-257
efficiencies 224-247
eggs/eggmobiles 50, 113, 117, 166, 171, 234-236
elder-care farms 296-298
electric fence 15, 118, 120, 209-120, 212
electronic aggregation 69, 135-139
emotional equity 89-90
emotional support .... 106-107, also see support network
entitlement 64
environmentalism 5, 9
erosion 41, 189
errand planning 239-244
everyone food 162-163, 267-268
excavation 13-14
exotic foods 162-163
exploitation of resources 4
extrovert .... see introvert

factory farming/livestock 6, 162-163, 296
farm cooperatives 129
farm definition 2-4
farm equity 220-221
farm shop 18
FarmER 3-4, 203-204

Index

Farmer's Progress	229
farmers markets	132-135
Farming Ladder	229
FarmLAND	3-4
fear	65-71
Featherman Poultry Processing Equipment	216
fecal soup	156
feed the world mentality	94-95, 102, 290
feedlots	6, also see factory farming
fences	14-16, 178, 187, 272
fiefdom	86-87, 222
firewood	46, 47, 49, 281
Food Inc.	24
food allergies	296
food co-ops	131-132
food police	140
food trucks	148-149, 180
Foreman, Pat	236
Fortier, Jean Martin	32, 205
Fried, Jason	157
frugality	10-11, 36-55, 79-80, 280-282
Gateway Products	165-166, 184
Genetically Modified Organisms (GMO)	296
germination tray	87, 106
Gladwell, Malcom *Outliers*	28, 63
Gobbledygos	119, 211-212
Godin, Seth	154, also see *Purple Cow*
Grasstain Tours	87, 288
greenhouses	117, 192, 196, 212, 236, 275
gross margins	88, 168-185, 199, 232, 234, 236, 300
guilds, historical	63, 256-257

311

Hartman, Ben 227, 242
Harvey, Paul 59, 163
healing the land 7-8, 11, 154, 188-190, 276, 305
Heinemeier Hansson, David 157
Henderson, George.. 32, 229
heritage genetics..266-268, 162-163, 292
Holmgren, Dave 16, 33, 186
honor box 143, 145
hoop houses... see greenhouses
horse 264-265
housing.. see domicile
Howard, Albert.. 6, 23, 32

Iberico ham 165
industrial farming.....................................5-6, 191, also see factory farms
infrastructure .... see portable infrastructure
instant gratification.. 258
integrity food movement... 24
introvert vs extrovert.. 80-82, 144
irrigation..120, 178-180, 204, 242

Jamestown, colony 189-190
Jeavons, John 204
Jefferson, Thomas 25, 28

Kelly, Kevin 210
Knowledge Society 22, 70

labor cost 54, 68, 83, 216-217, 218-223, 232-234, 236-237
land abandonment ....see abandonment
The Lean Farm... 227
"leave no trace"..7, also see abandonment
A Life Unburdened 157
lists, importance of..244-245

living on site ... 10
Lopez, Tai ... 22, 70
lunatic tours ... 2, 81, 124, 276

magazines ... 30
margins ... 181-185, 234, 300, also see gross margins
The Market Gardener ... 205
marketing ... 68, 92-107, 126-149, 150-167, 300
marriage/partnerships ... 76-77, 83-84
Martenson, Chris ... 63
masseuse of the land ... 9
Memorandums of Understanding (MOUs) ... 86, 88, 236
Metro-farm ... 153
Metropolitan Buying Clubs (MBCs) ... 83, 145-148
micromanagement ... 84-85
mid-life crisis ... 248-253
mid-lifers ... 248-253
middlemen ... 93, 99, 101, 103, 149
Millenium Feathernet ... 118, 212
mobile farms ... see portable infrastructure
modular farms ... 213-216, also see portable infrastructure
Mollison, Bill ... 16, 33, 186
monogrammed shirts ... 46, 80
mosaic farming/landscapes ... 187-189
multi-enterprise ... see stacking

Nation, Allan ... 22, 23, 27, 32, 163, 187, 249, 251, 258
nature deficit disorder ... 287
New Rules for the New Economy ... 210
newspapers ... 29-30
non-GMO feed ... 242-243

olive production ... 206-207
Olsen, Michael ... 153

on-farm store 143-145
Omnivore's Dilemma 24
open door policy 82, 145
Opportunities 163, 193-194, 284-302
ordinances 292, 144, 292, 295

Parsons, Stan 32, 90, 101, 181
participatory environmntalism 9, 19
partnerships .... see marriage/partnerships
Peak Prosperity 63
peer dependency 43, 45, 97-99
performance based remuneration
.... 83, also see commision based pay, fiefdoms
permaculture 16, 33, 94, 186, 196
pigs 35, 123, 165, 185, 195-196, 199, 219-220, 267, 270-271, 286
Plant-In-A-Box (PIB) 216
Pollan, Michael 33
Polyethylene pipe/PVC 16, 274, 279-280
ponds .... 13-17, 122, 178, 242, also see beaver ponds, irrigation
portable infrastructure 10, 66, 105, 202-223
post hole digging 227-229
post pounder 49-50
Pratt, Dave 52, 174, 181
price lists 156
pricing .... see gross margins
profitability .... 200, 278, also see gross margins
public speaking, fear of 75
Purple Cow 154, 267

quacker box 115

raken house (rabbit chicken) 116
Ramsey, Davd 59
Ranching for Profit schools 52, 90, 174, 181

raw milk 24, 141-142
reading, benefits of 22-23, 70, 255
recreation..265, also see vacation
ReGrarians 12, 33, 179
renting 48, 67, 177
restaurants ....139-141, also see chefs
retirement home .... see elder-care farms
reverting to wilderness ....see abandonment
Rework 157
road gutter 12-13
road system/road building...11-14

Salatin, Nellie... 38
sawmill 29-30, 67, 121, 208, 263, 295
school farms 299-300
shacks .... 10-11, also see domicile
shade cloth 211
Singing Frogs Farm... 32
situation awareness 22-34, 220
small supermarkets..130-131
Smith, Happy 38
Smithsonian research/studies 188-189
social media 137-139, 145, 154, 269, 271-273
spiders 188-189
stacking 67, 186-200, 206
starting to farm 248-261
stereotypes (of famers)... 35, 90, 289
stewardship 4-5, 7, 19, 104, 217-218, 265, 277
Stockman Grass Farmer ....22, 23, 163, see also Nation, Allan
StrengthsFinder Series 74
subcontractors ....68, 86-88, 159-160, 222, also see fiefdom
Sunrise Farms, non-GMO feed 243
support network 59-60, 62

tag-alongs 230
Taggart, Adam 63
team (assembling) 72-91, 232-233
therapy farms 293-296
too much choice 157-159, 165-166
tools 49, 53, 79, 88, 204, 205, 227, 240, 242, 273-275, 280-281, 290
The Tracker 34
tractor costs 41, 49, 54, 79, 204, 206, 231, 275, 281
Tribe 30-33, 59-60
Turkeys 112, 118-119, 162-163, 184, 186-187, 211-212, 266-268

urban farming 289-290
USDA 23, 194, 217

vacations 51, 264-266
value adding .... 103, also see stacking
vegetation cutting 19-20
Venezuela 39, 40, 168
virtual (farmers) markets 135-139
Voisin, Andre 32, 237-238

Wal-mart 104, 126-127, 132, 133, 135, 139, 281
water .... 16-17, 204, also see ponds
water bars 11
wealth generation .... 51, also see frugality
Weston A. Price Foundation (WAPF) 23, 157
wilderness .... 2-3, 5, 189, also see abandonment
Williams, Bud 51
Working In The Business (WITB) 181
Working On The Business (WOTB) 181, 245
Wounded Warrior Project 294

young professionals 254-256